THE FIFTH FLOOR

The Fifth Floor

A movie script of International scientists in the US

Andree Shalabi and Xiaoyan Wu

Copyright © 2023 by Andree Shalabi and Xiaoyan Wu.

Library of Congress Control Number:		2023916742
ISBN:	Softcover	979-8-3694-0659-5
	eBook	979-8-3694-0660-1

All rights reserved. No part of this book may be reproduced or transmitted in any form or by any means, electronic or mechanical, including photocopying, recording, or by any information storage and retrieval system, without permission in writing from the copyright owner.

Any people depicted in stock imagery provided by Getty Images are models, and such images are being used for illustrative purposes only. Certain stock imagery © Getty Images.

Print information available on the last page.

Rev. date: 10/20/2023

To order additional copies of this book, contact:
Xlibris
844-714-8691
www.Xlibris.com
Orders@Xlibris.com
855272

Based on a true story.

This is dedicated to our friendship

CONTENTS

Preface ix

How it started 1
Hunter 7
Bill 81
Joscha 92
Yanhong 104
Liudmila 146
Naomi 155
Edward 160
James 168
Vilma 176
Betty 185
Pam 186
Raj 187
Katarina 190
Jingle 193
Theodore, Edgar, Olivia 195
William 199
Amelia 202

David 204
Charles 206
About 20 years later 211

Glossary of names- The "who is who" 231

PREFACE

On one of the numerous occasions Xiaoyan and I were reminiscing about our stories in Chicago. We had known each other for many years by this time, having worked together at the University of Chicago Hospitals. We two, along with others we met there, really had become as close as friends could. We had shared all of the ups and downs possible in life, and we remained proud of each other's achievements and successes. And Xiaoyan had an idea. "What if," she said, "we wrote a book about these stories?" GREAT IDEA! By then, our good friend, Hunter, had become very sick, and that was a good incentive to get started. But how to do it? We picked one of the characters, Lilly, to be the narrator and to lead through the story. And we needed help from others. As a visiting scientist, Lilly had not been around constantly. Before Lilly settled in Chicago and got her own apartment she shared her studies with her home university in Switzerland. Not enough she had been sent for 3 months of every year abroad to far countries to collect poisonous material as her research focus was on toxicology. That meant, sometimes she needed to move on her return as it was too expensive to keep the room in her absence. As a consequence, she got to know everybody, in the lab and even at their homes, which would make a perfect narrator. We began to contact everyone to collect their stories. Hunter was interviewed via Skype as often as possible and all our other friends amazed us with their responses. In May 2012 I took a flight to NY, where Xiaoyan lived by then, to finalize our draft. In a mere six months, we finished most of the book. Xiaoyan's mother, Manli Yang, a known artist, kindly supported our project by providing several drawings. However, it took us another couple of years to finalize it as both of us had been too busy working. And, though some of us no longer live in Chicago or sadly, have passed away, this time we spent together we won't forget. Let's the movie begin!

HOW IT STARTED

In 1995, as a student of life sciences and medicine focusing on toxicology and neurosurgery, I went to work for a short time in the laboratory of Professor Zak for rotations. Unlike in the US, we would never address our professors by first name, but by title and last name. Professor Zak ran a good-sized lab in the Life Science Center at the University of Basel, Switzerland. I had him in classes before and was one of the very few who got good marks from him. And he had accepted me in his lab. My friend Radu, who had been studying chemistry at the University of Basel, nicknamed him "Zeus," after the head of the Greek gods who tossed thunderbolts when angry. Acceptance into the highest god's empire was quite an honor. By then, summer was coming, and I was looking for another lab. We students would have a break, and instead of spending the time vacationing like others I wanted to go abroad and work. So I went to ask Zeus about other labs abroad, where I could work as a summer intern. Soon, he returned with a list of three names. All of these labs were in the US. Initially I had not intended to travel that far, but I started by emailing the first one, which was in Chicago. While still typing my email to the second address, I already got a reply from the Chicago lab: I could come to work in Chicago, if I would stay at least three months. YESSS!

In the following weeks, I learned further details. The lab in Chicago was at the University of Chicago (U of C) and run by Lynn, a well-known tenured university professor and Lynn intended to have me help one of her graduate students, with whom I could also stay. So far, everything sounded fine. And I went to Chicago.

On arriving in Chicago, the first thing I learned was that I could not stay with the graduate student as planned. So, I immediately started looking for a place to live and a Chinese graduate student in Lynn's lab, Shuang, helped me. When I went to look at a room Shuang had seen advertized, I found two apartment buildings, each with about twelve apartments. All of the tenants were Chinese. It turned out that the advertisement had been in Chinese, too, and kind Shuang had even accompanied me to check on it. The lady subletting the room was about 35 years old. Like most other Chinese there, she spoke hardly any English. I had hoped to improve my English during my stay in Chicago, but that wouldn't happen here. The apartments were on 61st Street and Shuang told me to be careful. Everything south of 60th Street was considered unsafe at that time. Later, the U of C made an effort to attract university members to settle there by offering nice mortgages. However, because neighborhoods never end with one exact street, it is always wise to be careful. Still, I needed some place to stay quickly and the Chinese lady seemed nice, so I took the room ready to work in Lynn's lab.

However, the graduate student, I had been assigned to work with and I did not form a team. I was completely new to that field and needed clear instructions and explanations. But there was no time for it and I was expected to fully function instead. That was not going to work, and when I messed up the PhD student became furious and stormed off to see Lynn, leaving me with the other lab members, who had witnessed the scene. That was when I became good friends with Shuang and her friend Ning, another graduate student in the lab, who offered that I work with them. After a while, Lynn called me in and we talked. Lynn was very kind and wise. She said that, the graduate student, I was supposed to work with and I were very unlikely to go on working together. She also said that if I stayed in the lab but working for somebody else, we would still have to see each other. Therefore, she suggested that I move to another lab. Lynn arranged for me to work in Tali's lab, which was further up on the 3rd floor.

Tali, an associate professor, who had gotten her PhD from Tel Aviv University, ran a rather small lab with only three other people: Abigail, a postdoc; Alison, a graduate student; and an Indian technician with the last name Put — everybody had fun teasing us by combining my first name, Lilly, with her last name whenever they saw us together. Tali was married to Lior, another professor at the U of C and a graduate from the Weizmann Institute in Israel. They had two boys: a "little one" in elementary school and a teenager heading for college.

Tali gave me a completely new project, which was to demonstrate a circadian rhythm in the breathing of yeast. To be able to make measurements, we had to borrow equipment from all over the campus. I remember walking with Tali to the zoology department, among other places, until we collected the instruments we needed. Tali had a nice idea, but could that be all done in a mere 3 months? The other lab members expressed doubts that it would work. This was not really encouraging for me, especially after the experience in Lynn's lab and the project indeed got never really started. But during my short stay in her lab, I really appreciated Tali's kindness. She worked hard on all fronts, on her roles both in the lab and in her family. Her husband Lior, who ran a lab of himself, stepped by occasionally talking with a thick Hebrew accent, at times hard to understand. We talked about Chanukka, Tali being always concerned to drink enough, wondering, how aircrafts fly and other topics, making the lab a welcome chatty place.

HUNTER

In the middle of that, before the yeast project started, Hunter stepped in. Hunter, a tenured professor from the fifth floor resembling Robin Williams, was good friends with Lior and Tali. He told me he was British — from Wales, to be precise asking me about the whereabouts. When he heard what we were trying to achieve, e.g., checking for a circadian rhythm in the breathing of yeast, he immediately saw his chance and tried to convince me to work for him, instead. Hunter was working with tissue cultures, which was always good to learn. He was all about science and seemed to me a genius. So he took me up to the fifth floor and showed me his lab. I had cultured some tissues back in Basel, and I asked him if they would have what we called a "clean bench" for sterile work, a workplace necessary for tissue cultures. Hunter looked puzzled.

Suddenly, he wiped one of the lab tables in front of us and asked: "Now - is THAT clean enough?" When I explained what I meant, he said: "Oh, you mean a *hood*!" Of course, they had a hood, and I kind of liked this lab. Though I had personal reasons for not wanting to leave Tali, who had been so kind to take me in and was trying to establish this project despite of all rejections, Hunter's lab seemed a better choice, so I finally agreed to work for him. As a Swiss proverb says, it takes three times to get it right.

Hunter was kind of like Zeus. He was known to have quite a temper at times, but he was always very nice to me. And, of all my university professors, he was the one I learned most from. Hunter showed me everything. He showed me how to do tissue cultures with fibroblasts, PC12 cells, and astrocytes — which, to his amusement at my struggles pronouncing it correctly, I once called "asteroids". He showed me how to do immunohistochemistry staining and three-dimensonal confocal microscopy. His mind was bright like no one else's. Friends I knew would say that if anyone there would get the Nobel prize, then it would be Hunter. During seminars he attended with us, he would ask the right questions, and many of us admired him for that. He also did all math calculations without a calculator. Further, he did radioactive labeling himself, as it was cheaper. Money was a big issue for Hunter, and wherever he could save some, he would. He was also brilliant about movies, literature, and classical music. When he gave me rides home after work, mostly after midnight, he would turn the radio on and say, "I will get you home, if you identify the next piece!" Of course, I hardly identified anything, despite having some background in classical music, so we would keep circling and circling and talked about everything.

Once at home, I would go to sleep, but that never lasted long. His voice, speaking on my answering machine, would wake me. "Hey, Lilly! Are you there? Pick up! I know you are there!" When I picked up the phone, he would ask: "So, what are you doing?" "Hunter, I'm trying to get to *sleep*." And he said: "Oh. Well, *I* can't sleep." In hindsight, I don't know how I managed those days of getting home at around 1 am, talking to Hunter on the phone, and returning to the lab at around 7 am. Hunter, a night owl, rarely showed up before 11 am or later. But it was a great time!

So, I would get to the hospital at around 7 am the next morning. Usually, I would find Cooper there at that time. Cooper, a technician working for William, another professor with a lab on the fifth floor, was a young guy in his twenties who liked playing a computer game involving, paradoxically, *killing* people. One would think that everybody working in a hospital would focus on helping cure people, not trying to kill them. As part of a morning ritual, I would start chatting with him by asking him how many people he had killed so far. I remember once when Cooper complained of receiving a letter summoning him as a juror in a court trial. Unlike back to my home country, when people in the US get such invitations, they must attend. Whereas Cooper was rather annoyed about it, I would have very much liked to go. But of course, as I was not an American citizen, I would never get asked. We would occasionally talk about other things as well, since we were the first to arrive in the morning. Then I would quickly check my emails, replying Radu and continue working on whatever tasks remained from the day before. Around noon, Hunter would arrive. He would check on results, explaining and discussing further procedures. Hunter made up for what the graduate student in Lynn's lab had left out — explaining things to me. No other professor ever spent so much time training me. In between, he would scribble down more instructions to me. Sometimes I found these hard to read, but I quickly got used to his handwriting. Hunter checked on the other lab members, too, or he would work in his office with the door always locked, requiring us to knock if we needed to talk to him. At around 6 pm, he would disappear to eat supper and play squash. He was quite good at squash, and he would play it to exhaustion. One day, he came back quite excited. "Ha, I beat this guy — he's now finished!" And he meant it literally as the other guy broke his arm in a wild match – he was indeed finished! And Hunter won. Of course. At around 8 pm, Hunter would come back, and then we would work past midnight. That was our regular schedule.

He indeed had so much energy! While other professors like Bill would spend most of the time in his office, Hunter was always "hands on". He was very much present, seemed to supervise everything and while others felt sometimes he was too controlling, I enjoyed his presence. When he was away, we could just call him, there were no "stupid questions" and he simply gave advice or explained things to us. He would assign us to one task after the other, not to all at once, so we did not get overwhelmed. And there was no toxic micromanagement I had to face when I worked once for a lady professor in Heidelberg. For me, it was perfect leadership skills and he infected me with his passion for science right away. But as mentioned, Hunter used to have quite a temper. If some experiments didn't work, lab members were too afraid to tell him. Instead, they went to tell me and asked me, if I could drop it to him, so they were spared of his reaction. When I went to tell him, it was quite ok as I had not performed those experiments and it worked out well for all of us.

One day Hunter was quite annoyed, and I asked him why. He had gotten a couple of parking tickets and he had not paid them so far. He didn't want to bother with something like that and paying money was always an issue for him, especially for something he had no use of. So now he was ordered to court about. He was in a bad mood, but everything turned out ok later. We just didn't have to ask about.

My Chinese home, meanwhile, was the best address I could have picked. Yes, shootings occurred in the neighborhood almost every weekend, and a police officer would ring our door bell almost every Monday's morning to ask if we had seen anything. Hunter's graduate student Matthew, who lived few houses away at the corner of Ingleside, occasionally commented about these events. And living there did not help me to improve my English. But I learned all sorts of other things. Besides a few words of Chinese, I learned many things about Chinese customs. My landlady had left her little child with her husband and parents back in China while she tried to raise some money here in the US. She had asked me only one question when I first inquired about renting the room — and that was about my Chinese zodiac sign! I did not even know my Chinese zodiac sign at that time, but we figured it out quickly and by translation of Shuang I said I was a Sheep. Obviously that was not so bad, because she accepted me. And I lived like in another new world. One day, a camera crew from a Chinese TV station came to visit us to broadcast about Chinese living in the US. They also interviewed me, which would be broadcasted in China. On another occasion, my landlady and our neighbors all prepared for the Chinese Moon Festival. My landlady was teased for having prepared chicken feet — chopped off feet with the claws still attached — which I had never seen served before and which gave me the creeps. Then, yet another time, I was there as a young couple was about to get married. Again, the Chinese zodiac came into play. The young groom-to-be, born in the sign of the Sheep, announced that he wouldn't marry the prospective bride because she, as a Tiger, would probably kill him. And, I had to keep quiet about being a Sheep, too, in order not to be married off to the leftover Sheep…

On another occasion, two couples asked me if I wanted to join them to see the Indian summer in Canada. I had never heard of Indian summer, so I was happy to come along. We rented a car, and the five of us left for the weekend. Of course, there was not much verbal communication with me, since I didn't speak Mandarin or Cantonese and they didn't speak English, but all of us had so much fun! Everyone except me wanted to drive the car. It was like trying a new toy. Some of them drove *very* slowly, and I worried a little that the police would pull us over for that. From time to time we would get out, take pictures, and continue to drive. Whenever we stopped to use a restroom, I got excited and would always look under the toilet seats; toxicologists say that black widow spiders liked to hide out there. To my disappointment I never saw any. But when I went to do my laundry one morning, I backed up from the door of the laundry room because I saw a bright red spider with white dots, like nature's warning sign.

In Canada, we didn't have any room reservations and since it was the season of Indian summer with many tourists around we found it hard to find anything to stay. So we had to take a room with five beds. No problem for the Chinese! Something I would have never done back home, sharing a room with four neighbors I hardly knew, but here it was possible. We had breakfast at an American place next morning, and the waitress turned to me to take everybody's order. It was a funny situation, as I could not function as the expected translator despite obviously belonging to the group. Our waitress was really puzzled. But even without language skills our group managed just fine.

My first three months in Chicago passed all too quickly and the summer was over. One of the couples from the trip to Canada took me to the airport for my return to Switzerland. When my flight was delayed, they took pleasure in teaching me how to play Chinese chess, which is far more complicated than the Western version. I knew that I would miss these people, the lab, and the fifth floor. I decided to do everything I could to return.

So on arrival back home I spoke to the dean of my department and we worked out a plan allowing me to continue in Chicago. As I was not a US citizen Hunter would have to pay a huge amount of money to have me working, and to my total surprise he had even instantly agreed to do so! Together with the dean of my department we worked out another plan, which not only allowed me to continue working in

Chicago but also to save Hunter money with the right visa. Over the next years, I continued my work in Chicago, which had become home to me. Of course I had to present results at the university of my home country and I had to fly back from time to time. This way, I could always see my best friends Radu and Liesel. On my return, Hunter would write a letter to the US immigration agency each time saying "how important" I was for the lab, which made me always feel funny. Not only that, he was ready to argue with them in case there would be any problem. He also wrote letters to my university department head in my home county, to make sure I would earn all possible credit for my research in Chicago. I still keep his letters today.

Chicago really became my emotional home: beautiful Hyde Park, with its gothic architecture; big grey squirrels in fall; and a sky so blue that it looked almost painted. Whenever I returned at O'Hare airport, I could see the skyscrapers lined up along the water as the aircraft prepared to land. Then the plane would taxi along a narrow bridge right over the highway. Finally, we passengers would all walk the long hallway accompanied by flashing lights in all colors flying ahead of us. It was like directing me home.

Usually Hunter picked me up at the Airport, and he would sing one of the Beatles songs aloud, making me laugh. Once, as we approached Hyde Park, Hunter started yelling "Oh shit!" He had forgotten to check on the gas. We both pictured ourselves already hitchhiking, but we just made it back to Hyde Park. Hunter had a nice, sporty car that he would loan to Matthew, his graduate student, whenever he was away.

Soon, he started offering me his car when he had to leave town. But, lacking driving experience in the US, and afraid of being obliged to pick him up but not knowing how, I refused. Hunter took it somewhat as a rejection and responded, "My car is not my castle!"

Thus, I decided it was about time to get an Illinois driver's license. Unlike in Switzerland, a US driver's license doesn't require taking classes, and the best of all: it only cost about twenty dollars that time! Once the applicant has passed the written test, which is free, he or she can practice with a friend who does have a license.

All my friends were busy working. Shuang from Lynn's lab kindly offered to practice with me, but it was easier to call up a driving school as we didn't have a car. And I did not want to practice with Hunter's car! After only three hours of instruction, I headed to the testing place and passed the test. It was like heaven!

Then, Christmas was coming up. Being Jewish, this was not such a big deal for me, but Hunter asked me what I was doing. When I told him I had no plans, he asked me to come along with him to visit Giorgia and Claudio. They came from Italy to work as postdocs for William and had an eight-year-old son named Angelo. As typical Italians, they adored their son. Hunter always thought their son got spoiled too much. Angelo had an entire room full of toys, something, Hunter has never had. So we went for Christmas dinner to Giorgia and Claudio. Because the other guests were all couples, it felt a little bit strange coming with Hunter, but he had very kindly invited me along. Although dinner was fine, Hunter didn't like a chap called Leonardo, they started with disagreements and it was good when it came time to leave.

In a foreign country, everything seems so new. I still remember the first time I went to the post office on 61st Street. Except for the Chinese, the neighborhood was mainly African-American. I seemed like from another planet. So I waited in line like all the others; when it was my turn, I was amazed that they actually understood me and gave me what I ordered. I don't know what I expected, but I felt like a little girl who had bought her first stamps ever. And I had the same experience in the cafeteria downstairs from the lab. Almost none of us had a lot of money so we couldn't afford much. But it always felt like an adventure to go downstairs and spend a dollar on some all-new-to-me cornbread or a serving of macaroni and cheese. I never knew how to say it correctly — "macaroni and cheese" or "cheese macaroni, please," but the huge African-American lady there always gave me a big smile and asked, "With extra crust?" "Oh yes, extra crust, please!" That sounded great, despite all of the saturated fatty acids it contained. I had succeeded at ordering, I had "made it in the US" and I felt like I won the lottery.

Of course, I was not used to the language with all its different accents and pronunciation in the beginning. Whenever I talked science with Hunter, I had no idea how to pronounce scientific terms correctly in English. As I struggled to say them right away, Hunter must have got the impression that I had no idea about the subject. So, in a desperate attempt to avoid looking stupid, I took to pronouncing all chemical agents and cells in my mother tongue.

And I wasn't the only one with adventures in the language. We had some Chinese technicians working on the fifth floor who even hardly spoke any English. They had studied very hard and passed the Test of English as a Foreign Language (TOEFL), but they were not able to communicate. One day, a contamination occurred in the tissue culture room, and one of the technicians, talking about fungus, announced: "Mushrooms! There are mushrooms in tissue culture!" Hunter immediately picked up on it and, amused, asked: "Are they champignons?"

All of the foreign employees working at the U of C had to pass the TOEFL, so I decided to take it, too — even though it was not a requirement for me. However, my friends considered passing this test challenging, and I didn't want to miss out on that, feeling privileged. So, one Saturday morning, I went to the test site in Chicago and just took it. To my surprise, I passed with more than 99%. Even so, I knew very well that Hunter was just too kind when he assured me that my English was fine, I still didn't know how to pronounce certain scientific terms.

Hunter would take me out for dinner, and occasionally we would drive downtown to one of the recommended restaurants. Unlike in Europe, there is no free seating and one has to wait to be seated. Not long after I started working with him, we went to a restaurant which was on TV that night. Some kind of show was introducing good restaurants to the public, and they actually broadcast our dinner. That was something new to me, my dinner being broadcast, and even Hunter felt intimidated. Sometimes, if we had leftovers, Hunter would bring them the next day in doggie bags for dinner. Despite his tight grip around money, he was always very generous when it came to restaurants. One day, Hunter found a $5 bill on the street, which made him extremely happy. The lab members instantly announced: Hunter's going out to dinner tonight!

As Hunter knew a lot about movies, we occasionally went downtown to see one. I still remember the first time. We went to a movie theater, got the tickets, and then entered the auditorium just as the movie was about to start. But Hunter didn't like it. So we went to the neighboring auditorium to see what was playing there and, after ten minutes, to yet another one, and so on. The next day, when people in the lab asked me which movie we had seen, I honestly answered "I don't know."

Hunter's lab, however, looked rather outdated. Hunter had brought some equipment from Britain including this huge iron centrifuge that one can only admire in museums these days. Some faded photographs from Hunter's students' days hung on the walls, and the lab did not appear to be best organized. However, everything seemed to work, Hunter being "hands on" and knowing every single device and agent, it was functioning quite well and while Zeus's lab in Basel had been very modern, Hunter's lab seemed to produce more papers for publication. As Hunter was about my size, he discouraged me from wearing shoes with high heels, so I wouldn't be taller than him and we laughed about. Even more, he preferred me have lighter hair; for the next years, I would dye my hair. So no high heels and dyed hair, which of course was not an official requirement, but it was possibly better for his mood and general atmosphere.

At times, Hunter had also to teach. He was meticulous in what to teach and how to present it, and it took months for him to assemble the very best material. I doubt that the students could appreciate it; students of that age have little idea what are the latest developments in science. But current content was not enough. Hunter was also concerned about how students reacted to his lectures. So he asked me to act as a sort of spy by sitting in his classes to overhear what students said about him. I found this role very amusing. Of course, the other students asked me which classes I was in and which ones I had already taken. When I told them that I worked for Hunter (work in a double sense here), they showed deep respect and curiosity. They all wanted to know more about how it was to work for Hunter, and I tried my best to invite them to come see for themselves. But his reputation as a shark seemed to scare them away.

Then an incident happened in my 61st Street neighborhood. I had always felt safe and would take the green line bus from 63rd on Cottage Grove to get downtown. By that time, I had met many people in the neighborhood who knew that I worked in the U of C hospitals and who would stop me for a little chat or even tea. One afternoon, I did not feel safe. I had the feeling of being followed. When I made a sudden turn and hid behind some trash cans, an unpleasant guy passed me murmuring "Where the hell did she go?" I stayed there another 40 minutes or so, shaking and scarcely breathing, before I had the courage to leave my hideout. When I told Hunter about it, he got quite upset. From then on, he accompanied me whenever he could home.

Another incident happened a few years later, when I lived on 55th Street and Blackstone in my own apartment. One night, I walked home carrying all my microscope slides so that I could label them appropriately at home. That was years of work and of course very precious not only to me personally, but also to the project. When I turned the corner behind the bank on 55th Street, I found myself in the middle of a street raid; a gang was robbing people of their belongings. Gathering all my courage to defend my slides no matter what, I stepped right into it, calling out loud "EXCUSE ME!" and walked on past. The gang was absolutely taken by surprise and fortunately they didn't go after me. Of course, I was very lucky, but I was also determined to defend my slides — at almost any cost.

Raids like this happened, but they happened on other campuses, too, and despite of few occasions I never felt unsafe on campus. It is as everywhere a question how someone dresses and moves beside to be in the wrong place at the wrong time. Dolores, Hunter's ex-girlfriend used to dress expensively and therefore drew more attention and got robbed a few times. But she would not change.

On one of my flights returning to Chicago I met Elijah, the son of one of the U of C nurses I knew from the neighborhood south of the U of C campus. An African-American in his twenties who sang in the Chicago Gospel Choir, Elijah had a very warm and kind character, and I would meet up with him a couple of times when I had some extra time. One Saturday, on my birthday, he took me to his church. As the only non-African-American, I stuck out like an alien, which left me feeling quite intimidated. But everybody was so very welcoming, and to my great surprise, Elijah got everybody standing at the end of the service and the entire church sang "Happy Birthday!" to me. It was the most beautiful birthday celebration I had ever received. By the end of that year, Elijah came to Switzerland to sing with the Chicago Gospel Choir. They started coming to perform there every December, and whenever they did, he and I would arrange to meet in Bern or Lausanne, when I was not in Chicago. He always invited me to their show for free and I went to see it two to three times.

Meanwhile, the life in Chicago went on. During his tenure, Hunter had seen several department chairmen over the years come and go. Later, his colleagues David and Bill became department chairman, too. Before Bill, though, Bruce, who was the boss of Amelia's friend Carol, all of them professors at the U of C, was chairman of the department. He was a nice fellow, a skinny cardiologist, and Hunter liked Bruce's time in office. His only major flaw was that he was not a born leader. Rather than leading forcefully, he would collect everybody's opinion and then make a decision midway through this process. He stayed as department chairman for many years.

One day while I was there, we got a new department chairman, Bodhi. Everybody got very busy organizing presentations and the introductions for his arrival, and some looked forward to this event with mixed feelings. I happened to be in the secretary's office on the third floor at Leesha's and Leroy's, both very nice African Americans, I sometimes went for coffee with. And I learned that the new chair had to postpone the event, since he had an urgent dentist appointment. Back at the lab, I told Hunter that the new head needed a dentist and would therefore postpone the event. Hunter looked at me in disbelief, as if I was a fortune teller, and he was even more surprised when he got a notification email postponing the event.

The new head came with the reputation of being another shark, one who always asked questions nobody wanted to answer. On the actual day of the event, we all went to the seminar room, and to my surprise the new head took his seat right next to me. As his table was broken, he asked me if he also could place his coffee on my table, which of course was fine. During the presentations, the new head surprised everybody by not asking a single question. Because of my toxicology background, afterwards they all asked me, "Lilly, what did you put in his coffee?"

The new department chairman was about forty and talked a lot to people as he was very social. He was fair, but Sam, the boss of Hunter's ex girlfriend Dolores, didn't get along with him. Later, Sam became chairman himself. Sam originally came from Canada and was best friends with Bill and Yanhong. He stayed two to three years as department chairman and was good at it. Although Hunter and he had their problems, they got over them.

Of course, there were other labs around. One was that of Adrian on the third floor. Adrian was a very nice guy originally from Romania and I liked to step by and chat with people there. Adrian had an Indian lady postdoc working for him, who gave me insight into "arranged marriages". As she had no time for dating and as it was her custom, she had to meet three prospective candidates sent by her family and to choose one of them. In the US one could really meet all nations with their own customs.

On the second floor was the lab of Peng, a professor from China, Joscha went sometimes to as Peng was on his PhD committee. Bodhi had his lab on the third floor, cooperating with the professors Ava and Lynn. Others were around and in the Knapp building, the fifth floor once had to move to temporarily due to renovation and we shared room with Santosh, a very nice guy from India, who ran the lab with his wife from Canada from 1999. While we were in the Knapp building, my friends Liesel and her cousin

Ursula came to visit. It was funny to have the two elderly ladies stepping by daily in the high tech labs talking to people *Genglish*, as Hunter called it, a mix of German and English.

For years the department was employing scientists with different backgrounds. Then, the department got totally restructured and Sam was now cut off. For us, the department was no longer what it used to be. At about this time, William left the U of C.

On special occasions, Hunter would take his lab to a restaurant downtown. At that time, Hunter had a student, Laura, working for him. Laura came from a wealthy Chinese family. She drove a Porsche, so Hunter thought it would be fun to make all of us to drive with her to a restaurant one night. But Porsches don't have much space, and I ended up squeezed in between Laura and Hunter; we even crammed most of the people of the lab into the backseat. This was the craziest drive and full of laughing people!

At one point, Hunter had another student, Ron, working for him. Ron was a nice, quiet Jewish guy from medical school. Ron's father was an MD, and we would sometimes talk a little bit about his lab project and life in general. Like Laura and so many other students he just stayed for a while.

On weekends, not many people came to the lab; on Sundays, Hunter and I would be the only ones working. Sundays at noon, Hunter would call his mother in Bristol. She was a very sweet old lady with a very British accent, and Hunter annoyed by his mother's conversation often handed me the phone to continue to talk to her. I had become used to American English quickly, but understanding an old lady in British English I was quite lost at times — not sure whether to pity or congratulate her on what she told. To avoid embarrassment I would just talk as much as I could myself and she might have wondered how come I was so chatty. We would keep in touch, and I would send her postcards from my travels and holiday cards. When I was abroad, his mother provided a reliable resource to help me keep in touch with Hunter if I had not heard from him a while since Hunter was not good at keeping in touch by writing and we didn't have had a service like skype yet back at that time. I would call her and, when the time came for her to talk to Hunter on the following Sunday, she would remind him to get back to me. And he always did.

One time, when I was returning to Chicago, my flight made the news. I was flying via Britain, and just after leaving Heathrow, we had an engine failure. An aircraft can continue on three engines, but another engine failure occurred just as we started to cross the Atlantic. Therefore, the pilots decided to return to Heathrow. As their protocol requires, they released their kerosene and we went back. But, as it turned out, they had released too much of the kerosene, forcing us into an emergency landing. Unfortunately, I was carrying a household of items on this flight, and, when we finally landed on the tarmac, we had to carry all of our own belongings across the airfield. On top of this, the newscast showed me amid all of my possessions. We were forced to stay overnight in a hotel at Heathrow airport in London. After I got to the hotel, I decided to give Hunter's mother a call, now being in Britain for a couple of hours, and she immediately asked, "Was that you, honey, with the suitcases, the chair, the lamp and all on TV?" Luckily, Joscha, one of our best friends at the U of C, had not asked me to bring back car replacement parts that time!

Hunter always flew on United Airlines. One time, we decided to travel together to Heathrow, where I would continue on to Switzerland and he to Bristol to visit his mother. Of course, we wanted to sit together, and I called United to arrange that. United assigned Hunter another aisle seat, but when we checked in, he became terribly upset that they had changed his seat. This just didn't make sense. As soon as we were seated, he fell into a sound sleep, showing how exhausted he actually was. But he woke up again soon and stayed up all night. Annoyed about a guy sitting in front of us, he would try to return "favours" to the guy as it happens when people have too much energy. At Heathrow, Hunter hurried like he was being paid to make the next bus to Bristol, while I still had some time to kill before my next flight. But, as always when he felt something was not right, Hunter apologized to me, and I couldn't remain angry at him.

The Fifth Floor

On another of my returns to Chicago, I brought an espresso machine. Hunter liked coffee, but American coffee tastes far more diluted than European coffee. The time seemed right for the lab to have a nice machine. Because the voltage in European wall plugs is higher than in the US, I even brought a transformer, which was pretty heavy. Our nice coffee times, however, didn't last long — the machine, including the transformer, got stolen. At times, stealing at the hospital was a big issue.

The third floor of our building had an unusually spacious lady's room, which I preferred over the one on the fifth floor. Once, when I was in the third-floor lavatory, I overheard Hunter talking to Tali as they walked down the hall. They stopped right in front of the lavatory door, and I hoped they would soon continue on. Instead, they fell into a disagreement; by then, having waited already a good fifteen minutes for them to leave, it would have looked awkward if I'd come out. Trapped, I was forced to listen to their entire discussion. It was another thirty minutes before I could escape. What else could I do?!

Hunter and I used to work with ^{125}I, a radioactive chemical element. My toxicology background made it quite normal for me. Others, like Hunter's student Betty, always feared working with radioactive materials. Hunter labeled transferrin with ^{125}I, and then we would do uptake experiments under a time course, Hunter adding the hot stuff and I removing it with a glass pipette by a vacuum device. We made an ideal team, and Hunter always seemed to find the results very exciting. Standing in front of the gamma counter with me as the results came up, he would snatch the printout before I could, calling out loud: "Wow, this is better than sex!" One day, however, some of the added radioactive solution got on my glove. In no time, Hunter cleaned it off, checking me with a gamma counter afterwards. The device didn't detect anything, and I was lucky not to get sent into quarantine or worse.

Weather in Chicago could be dramatic, with a wide temperature range running from +40 °C in the summer to -40 °C in the winter. It experienced heavy, heavy thunderstorms and the most beautiful sunsets, which I liked to watch from

Hunter's lab overlooking the Knapp building to the west. One night at the lab, while watching the sky, I noticed some unusual activity down on the street by our local post office. In fact, some sort of burglary was taking place. Luckily, I had not turned the lights on, which would have made me visible to the thieves outside. While I was able to watch the remaining activity I could report it to police.

I always enjoyed living in Hyde Park very much. The U of C in Hyde Park has been described as "where the fun ends," but this was not true! Most campuses in Europe don't have as much dignified flair and beauty as the quadrangle or the Hutchinson Commons dining room. And, unlike so many other places, Hyde Park has huge, beautiful trees to look at while listening to the bright green parakeets flying in between. These birds build tremendous nests, each of which accommodates about thirty birds, and sometimes one can see them on telephone poles like the one on Woodlawn. On the two big trees across from my place on 55th Street and Blackstone, where I rented my own apartment, hundreds of different birds would assemble every night. It was loud as hell, but what a great site for bird watching! My building and neighborhood also offered other entertainment, so I never needed to buy a TV.

I had one neighbor to the right, named TC. I still don't really know how to spell his name, but because it sounded exactly like the two letters, I always wrote it that way. TC, a huge African-American guy with a great sense of humor, worked in Cook County Hospital as a nurse. When his family sometimes came to visit, it was great fun to watch them with their many long silver chains around their necks and a huge radio playing on their shoulder. Once, when a TV station aired a report about one of the U of C labs that I really wanted to see, I asked TC if I could watch it at his place since I didn't have a TV. He was very kind and invited me over, but unfortunately we couldn't find the right channel. Whenever I had to fly abroad in between, I let TC know, and he promised to keep an eye on my apartment. Marc, another tenant in my building, was a short guy who worked at the reception of our building. Other tenants included April and her husband, both very nice janitors, and Roman, who worked in the department of social science with

James, another of our friends. Roman took care of my mail whenever I was away and we would chat as soon as I was back picking up my mail. One big advantage of living in Hyde Park was that most people living there worked for the U of C and it felt like a big family.

But my building had still more to offer. One evening, the police stopped a car just in front of my window. While sitting in my kitchen, I had a front-row seat of some arrests. And 55th Street wasn't even considered a not-so-safe neighborhood. Now, they sometimes block parts of Hyde Park a little further north whenever President Obama comes to town — as his house is in Hyde Park, too.

During the week, we would attend seminars and sometimes company product presentations. We also went to libraries — which, unlike the ones at my home university, were open all the time, including weekends. A real dream for scientists! I always liked to go over to the John Crear library -a name none of us foreigners ever knew how to pronounce correctly- to spend time looking at newly arrived journals.

Some weekends, I would go to yard sales

— more to see how Americans live than to buy anything. I always found these sales so interesting. Where I come from, things are built to last for the next hundred years. But here, one could see bath tubs made of some kind of plastic which would squeak when being used, the practical built-in cabinets that we Europeans envy Americans for, and fridges that all seemed to be super-sized. In some places, I would even see a cockroach, but this could also happen in the hospitals (in a double sense, as Hunter would call people he didn't like cockroaches, too).

I would occasionally take the Green Line train, the "El" from Cottage Grove, which seemed like an acrobatic act when it rounded the curve at King Drive, announced by the conductor in a sonorous voice as "Kiiiing Drrriiiive!" Different people would board the train from better and worse neighborhoods, some reciting a text for begging. Most times I didn't understand a word of it. Riding the buses was another experience. One had to pull a long bell string to request a stop, which was a smarter solution than the randomly scattered stop buttons found in European buses.

I would sometimes ride the buses to Ford city, taking one from 55th to the Midway airport and another bus from there to the mall. My good friend Vilma had once introduced me to that area as her family was not living far away from it. The mall had a big Target store and a Home Depot, which had plants. Everybody was always very friendly when I carried my newly purchased household stuff home on the bus. For plants, I would get advice on how to water them by passengers; for furniture, I would often get helping hands. People in Chicago are really nice.

For shopping, the US seemed just like paradise. Everything was so much cheaper than back in Switzerland. Not only that, the variety of things was beyond imagination. The salespeople were very friendly and let children touch and try out toys in the store. Unbelievable! Once, when Joscha took his son Dominik to Hungary, little Dominik wasn't used to the European way and got shocked when the salesperson ordered him yellingly to back off from a toy in the store. Customer service was clearly an American invention, not a European one.

I would sometimes take the Green Line or Bus No. six, the Jeffrey Express to go downtown. There was the Chicago Art Museum or my favorite, the Field Museum, one of the nicest science museums I know. The entrance fee was rather expensive, so I could only go on rare occasions. There were the big department stores like Macy's or Marshall Fields, which had beautiful decorations at Christmas time, which were worthwhile to see. Or I would go to Filene's Basement to shop for clothes or to World Market further north, often just to look. One could find all possible sizes of clothes, from super skinny to super fat, and it was so much fun to find something suitable and not even very expensive. But everything was designed for "use it or lose it," not to change or to fix. I once made the mistake of cutting all the buttons off a dress, as I wanted to buy nicer ones. But, unlike in Switzerland, I couldn't find any to buy, and it didn't matter in the end as I never wore that piece of clothes. In December, Yanhong and I would go Christmas shopping together to Fox Valley. Her brother lived about an hour's drive away in Fox Valley - that is why Yanhong would know this area and trunk loaded after we had spent at least half of the day we returned home.

When I lived on 55th Street and Blackstone, Liesel from Basel and her cousin Ursula from Amsterdam came to visit. I had met Liesel in 1988 and we had been best friends since then. Liesel like my friend Radu lived in Basel and she was like an aunt for me. I also had met her cousin Ursula from Amsterdam on several occasions in Basel and her home town. Both were close to Hunter's mom's age, between 70 and 80. I simply offered them my apartment, while I took refuge at Vilma's place. Vilma worked as a student in Hunter's lab at that time and was not only one of the first people I met in Chicago but also one of my best friends.

Liesel, born 1928, had been in a relationship with Erich, a Jewish guy from Germany, who had once opened the first department store in Tel Aviv, where the Diezengoff Center is located today. That happened many years back around 1937 and at the same place in 1971 or so, Hunter and Ilana were offered a store. And, another good 30 years later, Dan and I would meet for coffee there whenever I was in Israel.

After World War II, Liesel visited Erich in Israel on several occasions. At that time, Israel was full of people who had survived the Nazi terror and were looking for jobs. It was not unusual for a person in a common job to be a former academic. Liesel recalls an incident where she ordered coffee with milk after the meal. The waiter, a former university professor from Germany, took the order with a paternal smile and said he would serve her the coffee with milk if they would take a seat on the veranda — so no one would notice that she had milk after her meal with meat, which was not kosher.

When Liesel visited me in Chicago, she was bringing my beloved violin, which I had left behind because I worried that it might get damaged while I carried many other things. The violin was the only memento left from my grandfather, who had studied music in Vienna and was killed by the Nazis. It was a typical orchestra violin, blackened as was customary in the early twentieth century for orchestra instruments to make them invisible in the orchestra pit. When I picked the two ladies up at O'Hare Airport, Liesel was accompanied, as it seemed, by admirers. It turned out that everybody had assumed she must be a famous violinist, traveling at her age with nothing other than a little suitcase and — actually MY — violin. The nonchalant Liesel did not correct anybody in this belief.

Liesel and Ursula visited the lab a couple of times, which by then had moved to the Knapp building to allow renovations, and met most of the lab members. After their visit, Hunter would always ask about her and the ladies would ask about him.

The ladies must have reminded Hunter on his mother. He always wanted her to come to visit in Chicago, and I offered her my apartment, too. Sadly, she never made it. He would visit her in Bristol once or twice a year, and I would occasionally talk to Hunter in Britain then. As old ladies do, his mother would sometimes require if I had a boyfriend. Years later, when I said yes, there was a long silence.

Hunter emailed me when she passed away and I no longer lived in Chicago, and I decided to attend the funeral in Bristol. Hunter showed me her kitchen, where she had set up all the pictures and postcards I had sent her over the years — "an altar," as Hunter put it. Hunter and Evelyn, his wife by then, came to pick me up at the airport, and Hunter waited in the car while she went into the terminal to find me. Hunter had described me to Evelyn as a girl with light-colored hair. Not working in Chicago any more, I had obtained my natural hair color back by then. He was pretty much amused to see me with my hair back in its darker natural color, joking that next time I should come like the Irish with red hair! They were driving his mom's old car; she had been still driving herself until recently. Amelia, Hunter's colleague, had sent a beautiful bouquet and Hunter was moved by how many people actually attended the funeral.

While I stayed in a hotel in Bristol, which was not only very expensive but also had a problem with moths, Hunter and Evelyn were busy with his mum's belongings. We took a very nice walk together and it was the first time I got to know Evelyn. She had worked as a school teacher and in the school library before. I thought she must be a very strong character.

Hunter's mom had been a school teacher, too. She was easy to get along with, but too weak to stand up to her domineering husband, also a school teacher. Hunter grew up in Sheffield, Yorkshire, where his parents had been sent to work as teachers at the end of the forties. Their salaries were very low. Hunter remembered having seen a wage slip, stating his mother's income as only one pound per week. While his father probably earned slightly more just for being a man, his parents lived at the low end of the middle class. They had no central heating, and Hunter remembered his bed was always freezing cold. For Christmas, Hunter would get three gifts, only one of them something fun for children, and that was all.

Their house was located in the outskirts of Sheffield, in a village surrounding. Their village had one shopping center, consisting of only three shops. For school, he had simply to walk down the street. His mother would cook him a meal for lunch. After lunch he was pretty much left to himself, as he had no siblings. Rather than playing with other children, Hunter would play cricket — which he loved — left hand against his right hand. He even kept score!

When Hunter was about twelve, they moved to Bristol, because his father had mistakenly thought to get better pay there. It was from 54 Cinder Hill Lane in Sheffield to 54 Evelyn Leaze in Bristol, where his parents bought a house. His father also bought a piano, so Hunter could start to play but never bothered to get Hunter lessons. Later, Hunter started to play some self-taught, free-style, avant-garde pieces on it with a friend, and his stern father would ask them to stop after a while.

During school, Hunter did not have it easy. Children made fun of his last name, and at school he was forced to eat rice pudding, which he hated to death. His father wanted Hunter to become a clerk at a bank, as he thought he had no talent to live on his own. A banking career would at least give his life stability. Hunter simply ran away and ended up on the street.

A year early he had applied to get into university, and therefore did not do so well. He applied at several universities with an A-level report, which must have been ten to twenty points or so in the British system, when someone suggested that he should apply to the botany program at University of Manchester, which would get him into university; he could switch to another subject later. So he did.

In Manchester, Hunter joined a jazz band, an "Avant-garde of Maniacs" as he calls them, and stayed for the next three years. While the other musicians were actually music students, Hunter was the only one from a different field. He had bought a trumpet for about £20, around $100 today, and he wanted to become a jazz trumpeter. He even tried to introduce some electronics (in the 60s, he was well ahead of his time) which, unfortunately, didn't work so well.

He was enrolled in botany then, but he didn't really enjoy it. What should he do? He had the chance to switch to the psychology program, as a friend of his did, but he would have needed to start all over and to spend four years instead of three. And he didn't want to do that. So he stayed in botany without much enthusiasm and went to concerts, movies etc. He only studied during the last couple of months before his finals. He got his BA with a major in botany and a minor in zoology, earning a 2/2. That didn't allow him to go on. One of his instructors told him to go to the US to apply for a PhD program. At that time, students would go from their BA straight to a PhD; a Master's degree was only considered for those who didn't make it in the PhD program. So he applied at Raleigh University in North Carolina and got accepted into a program where he would do tobacco research.

Since botany was not his love, he was not happy there either, but he stayed for two years. He had no interest in tobacco research at all. He didn't even really smoke at that time. Ronald Reagan appeared as an actor in advertisements promoting Chesterfields, the brand name of some really strong cigarettes. Later, Hunter would try cigarettes from his uncle in Britain, who smoked Senior Service, another brand of strong cigarettes, even unfiltered. His uncle would proudly smoke more than twenty a day. The Gitanes brand was considered the strongest and people would say, "This is the test!" Hunter started to like some strong oval cigarettes with Turkish tobacco. And that was pretty much the only interest Hunter had in tobacco.

Once, when he visited a poetry class, Hunter felt he should maybe leave science and go into humanities. He told this idea to his supervisor, the tobacco guy, who really got upset, since he had gotten Hunter in. But Hunter wouldn't stay.

In North Carolina, Hunter met and spent time with a large group of people. One of them, Alfredo, was an architect of some sort. Alfredo was of Italian descent and very adventurous; he introduced Hunter to a group of proto hippie types listening to bluegrass music. In 1969, Alfredo and Hunter went on a trip to Europe. They hitchhiked all over. They saw West and East Berlin and went down to Switzerland, where two nice girls gave them a lift. This made the guys quite excited. They went to lunch with the girls, but not long after the girls headed for the bathroom, they noticed that the girls had actually left without them. So Hunter and Alfredo were stranded at a rest stop in the nowhere of the Alps.

They continued on to Southern France and then to Spain — where they went to Barcelona,

Ibiza (which was a hippie haven) and Formentera. They basically slept on the beach. Then Hunter stepped onto a sea urchin. He went to see a doctor, who offered him two options: either to have the spurs removed, which would be very painful; or to let them grow out. Hunter chose the latter, but with that handicap, the trip was over for Hunter and he returned to Bristol while Alfredo went on to Morocco. Hunter would get postcards from Tangier, Marrakesh, or Casablanca proudly stating how much pot his friend was smoking there. After two months or so, Alfredo came to visit him in Bristol.

Hunter did not continue his studies in North Carolina and returned to Britain. At home, he really started to drift and got close to bad company. His father somehow found out that he had dropped his studies and problems arose. Hunter's father had always been very critical of Hunter and could be quite unpleasant. So Hunter went to London, really to start his "hippie time," as he called it. He had no money, but he had an old car, a van, he had bought second hand for £100 (about $500 today). The roomy van could carry a lot of things, which helped Hunter to make money by helping people move.

Back in North Carolina, students got pulled into the Vietnam war. One guy who was about to graduate shot himself in the foot to avoid being sent to war. Hunter, meanwhile, worked at a couple of odd jobs, among them selling vacuum cleaners (or at least trying to sell them) from door to door. People would be kind enough to let him in, but after he explained all one can know about "Hoovers," as they call them in Britain, people would say good-bye. They only wanted to have company, which still happens today mainly with elderly ladies.

He also worked for Derwant Application, a firm whose employees wrote agricultural abstracts. As one of about fifty employees, Hunter had to write about fifty abstracts per day. After he saved some money, he quit and went on a trip to Amsterdam. On the train, he met Ilana, a Jewish girl who ran a clothing business in London; she went to see her relatives in Amsterdam. On arrival, Hunter got a cheap place at a YMCA, sharing his room with nine others, and he stayed there for a couple of weeks. During that time, he would meet up with Ilana occasionally. When they returned to London, they got together. Then they headed to Amsterdam again.

They took one of the first hovercraft ferries and drove Hunter's van to Amsterdam. Not used to driving on the right-hand side of the road, Hunter had an accident. The van was in bad shape, and Hunter wanted to have the van fixed in Bristol. He put fiberglass all over and even painted it, but the wheel was damaged and he had to hold the steering wheel tightly all the time while driving. In the end, he gave the car to a friend. Hunter and Ilana opened a shop designing and selling clothes in London, and they got married.

Ilana, born and raised in Israel — a sabra, as Israelis call Israeli-born citizens — had parents from Amsterdam. She got Dutch citizenship and still had relatives in the Netherlands. Her grandmother on her mother's side, in her eighties, lived there, as did her aunt, a pianist. The aunt had wanted to settle in Israel, but she got stuck in Amsterdam and stayed. She was not an easy character and had little interest in Ilana and Hunter. Ilana also still had family in Israel. Her mother and stepfather lived in a beautiful house in Haifa on Carmel mountain, and she had other relatives in Israel, too. Her mother was a very curious person, and it was great to have conversations with her. Her stepfather was a left-wing theater director. Ilana's father was a life-long kibbutznik — according to Hunter, a real nice fellow but not intellectual. Ilana had a younger brother, who became an Internet service provider in Israel, and a younger sister. Later, Ilana's mother and stepfather would separate, and her stepfather, Hunter pitied on that occasion, tried to live on his own for a while.

The business in London went quite well, and Hunter learned to tailor clothes. But Ilana wanted to relocate to Israel again, and they gave up their business in London to open a new one in Tel Aviv. They moved into a big house close to a main road that had been turned into apartments. They were offered as mentioned a location for their business in the Diezengoff Center, Tel Aviv's famous shopping mall. But, because Israel was going through a period of money devaluation, they didn't buy that store. Further, they soon found that the clothing market in Israel was quite unlike the British one, and they struggled to make any money. After about nine months, Hunter decided to return to science in order to earn a living. He had the courage just to step in at a back door to the renowned Weizmann Institute. He told them that he hadn't done science for the last four to five years, but selling clothes instead, and could they, please, accept him as PhD student?" According to Hunter, he would have hardly had a chance to enter one of the top schools, but war was in the air, and Israel needed scientists. After a long discussion in Hebrew that he didn't understand, they said to him, "OK. Come back on Monday!" And that's how Hunter entered their PhD program. (Years later, I would just do the same, knock at a back door by email and kindly offered a position.)

This was not an easy time for him. The unloved botany followed Hunter, and for six months he worked with Gus on the transport of trichoderma. Gus was a young guy who worked with Hunter in the darkroom, and Hunter got interested in transport systems. Then Ariel came along. He had a background in microtubules and had spent a sabbatical at NIH, where he worked in neuroscience. He wanted to open a neurobiology Department at Weizmann in 1975 and needed people.

As Hunter really wanted to get out of the plant department, he transferred to neurobiology to work under a new principal investigator (PI), Ido. Unfortunately, Ido was neither an easygoing character nor a good teacher. He treated Hunter like a slave. Hunter's situation got worse when a guy from Amsterdam arrived; Ido then was no longer willing even to talk to Hunter. Instead, he left Hunter doing boring experiments, like measuring membrane potentials. As a consequence, Hunter got very fed up. When he saw Ido doing electrophysiology experiments, Hunter often felt the urge to give his PI a little push to finish him off. However, being treated that badly made Hunter an excellent teacher later, as he would be aware of where people would cease to understand, just as he had as a student.

Because Ariel was the department chairman, Hunter told him about his situation. Luckily, Hunter got a lot of help from Ariel, who took over control of Hunter's project — even though this created a bad situation with Ido, who became resentful. Ariel smoked a pipe, and whenever Hunter went to his office, he entered a cloud of smoke. The pipe seemed to calm Ariel, who was a friendly accessible character. Hunter spoke to Ariel a couple of years ago, after Ariel mentioned Hunter in something scientifically he published about his career in JBC. Another helpful PI who supported Hunter was Moshe. He made Hunter think independently. Hunter calls him a terrific teacher, perhaps his best. Moshe would come and help Hunter with his experiments. Hunter still remains in contact with him today.

While Hunter worked in the lab, his wife Ilana continued to design and sell clothes on her own. She became successful when she started to produce T-shirts for the Israeli market and worked with a guy called Gad, whom Hunter describes as an extremely nice guy. But Hunter's deep absorption in science meant that he spent most of his time now at the institute, while Ilana discovered the Israeli market. After only a year and a half in Israel, the couple divorced. Ilana met a new guy, who would become her common-law husband and the father of her two daughters. As Ilana kept Hunter's last name while the children were born, they inherited Hunter's last name.

Hunter lived alone now in a country where he didn't speak the language, but he stayed and carried on. He bought a car for $1000 so he could drive to the institute. When the war broke out, Hunter mostly worked alone in the lab, as his Israeli colleagues had to join the army. A guard would arrive at 5 pm sharp to ask him to switch off lights and leave, but it is hard to stop a busy scientist, and Hunter did as much as he could to continue a little further. Thanks to Ariel, and with the help of Moshe, Hunter completed his PhD, which came partly from his work with Ido and partly with Ariel.

After five years in the PhD program, Hunter went to Yale in 1976 as a postdoc and worked for Dr. Greengard**, who later had a big lab and was awarded the Nobel prize. Dr. Greengard has worked at Rockefeller University in New York (NY). He set Hunter to work on bird erythrocytes (red blood cells), which have nuclei, unlike human

---

* In order to recognize famous scientists, names are displayed as title plus genuine last names.

erythrocytes. Hunter's project was to investigate the Na-K-2Cl (NKCC2) transport system. Hunter would get turkey blood delivered twice a week, so he didn't even have to go to a slaughter house. He did radioactive uptakes of ^{32}P and studied the effects of Na-K-2Cl inhibitors. In addition, he worked on astrocolin, which did not interest many people at that time. A new fellow, Owen, joined the lab; he seemed uninterested in Hunter's topic, too. He was very secretive about his own activities, but he returned to Hunter's topic later and published more than one paper in the journal *Nature* after Hunter left Yale. With that, Owen's career took off; he holds a leading position at John Hopkins today. If Hunter had just stayed a little longer, at least one great paper would have been his.

At that time Hunter was dating a very nice girl from Santa Barbara, California; together, they published a paper on torpedo, a genus of electric rays, and electric eels. She works in Oregon today.

After Yale, Hunter went for interviews. One was in Göttingen, Germany, where he met Dr.

Neher, Nobel Prize winner in 1991, who worked there in single-channel recording. Hunter did not take the position in Göttingen. In 1984, he went to Chicago instead, taking a position as an assistant professor, later becoming an associate professor there and tenured.

Hunter said that in the mid-eighties, around 1987, when he worked as an assistant professor at the U of C, he got into EF2. Back then he should have pursued it, but it wasn't his top priority and therefore just couldn't go for it. Despite the importance of skill, passion for the subject is what really makes a good scientist. One of my professors I encountered on rotations in Basel had gone for something, which was not his passion and he kept counting the years until retirement. In science and all other areas it is not important, what one chooses, but we have to do it with passion.

By this time, Ilana had moved back to the Netherlands, living there as a fashion designer with her two teenage daughters. While I worked for Hunter, he went to visit them for the first time in many years. He joked that they could be his daughters. He had asked me to call him while he was there, and I did as he requested — probably making him look important and desired. Ilana and Hunter got along, but they managed to get into a fight about something as unimportant as vitamins, and the contact froze again.

What most people might have missed about Hunter was how much he actually cared about people. One day, he came into the lab and said that he had found a homeless man in his entryway that morning, and he "didn't have the guts to kick him out," because he remembered his own days on the street.

Another incident occurred when we got a new cleaning lady on the fifth floor. For many years, we had an elderly Polish guy who hardly spoke any English. One day, he suffered a heart attack right on our floor. One of William's people, Luan, who was working on his MD-PhD, found and rescued him. After that, we got a skinny fifty-year-old African-American lady from the neighborhood, Afra. Although Chicago's Hyde Park is beautiful, it is surrounded by not-so-safe neighborhoods. Our new cleaning lady was a former prostitute, and Hunter immediately commented on this: "I can't see how she made any money!" She lived with her underage daughter and her mother. This job meant a fortune to her, but sadly her mother demanded that she would stay home for her daughter. As a consequence, they fought constantly. When she did not show up for three days, Hunter got worried that something serious had happened to her. But Afra showed up again. The story was she had been jailed for driving without a driver's license! Every night she would arrive at around 8 pm, when the fifth floor was mostly empty. She dreamed of having another life, and I introduced her to Internet chat rooms, which I never used for myself. These sites let her write under the name "Princess", and no one would ever know her real identity. In return, Afra kindly offered to do easy tasks as turning off a switch, because we had to stay sometimes for hours just to do that. Everything was working well, when her situation at home deteriorated. Her mother kicked her out.

From then on, Afra had to live on the street. Now, as mentioned, the neighborhoods around Hyde Park like in all other big cities are not very safe, and a fifty-year-old woman on the street may actually not survive. So, Hunter took pity on her. We knew that Hunter's colleague Bill had a white sofa bed in his office. Bill always left at around 5 pm and returned early in the morning. As Afra had to return her keys after midnight, Hunter and I let her in again, and she would spend the night on Bill's sofa. When I came in around 7 am in the morning, I made sure she was awake and could leave. I would also share my dinner sandwiches with her. One of William's students, Virginia, even went so far as to bring in some bedding for her. Again, things were working out well. But the day came when Afra's family situation changed again, and our cleaning lady had to leave. To show her gratitude, Afra bought each of us — Viginia, Hunter, and me — one of the New Orleans Mardi Grass necklaces from a One Dollar store. She put them around our necks like in the Olympic games. Hunter and I were reluctant to also receive her hug as she had not have a chance to shower for a few days, and Hunter would grimace while I could hardly hold my laughter!

Hunter also had a romantic side. He liked the night sky, and he once told me that he had been always very fond of it. Back in his youth, the sky held dozens of visible stars; oddly, today we consider ourselves lucky just to see a few.

In the beginning of my time in Chicago, I lived in several different places, moving when I had to. Each time, Hunter would find a place for me. My first residence was in a room I sublet from the Chinese woman on 61st Street. When this became unavailable, because I had to return to Switzerland for a month, Hunter found me a place on 53rd Street and Blackstone. It was temporarily as the Indian guy living in this place just had filed divorce and was going back to India for a while, looking to sublet his place. So I had an entire apartment to myself, which was great! When I had to move again, about six to seven times, before I finally got my own apartment, Hunter let me put my belongings in his basement. I had a nice set of Chinese garden furniture and a Japanese shoji screen that I painted after spending eight hours getting the right color. People in the lab were always teasing me about "the right color," because I did not like to use tubes or any other equipment with colors that did not match.

Later, from the time, when I was working a side job for the US State Department, I befriended a couple working at the US consulate in Frankfurt. When I had transferred back to Switzerland again, they offered to bring things back from Chicago to Frankfurt. That way, I would not having to ship them, which was very expensive. And I contacted Hunter. This couple, originally from Chicago, was transferring from Frankfurt to their post in Bahrain. On the way, they planned fly home for their leave. They offered to pick up some of my belongings in Chicago and drop them off in Frankfurt. I asked if they could possibly pick up just the Chinese garden furniture and the Japanese shoji screen from Hunter's basement. So they showed up at the lab; Hunter, trusting them, simply gave them his keys. A couple of weeks later, I started to receive little parcels from Chicago. To my amazement, they contained all kind of things, like postcards, books, or scarves ... things I had not asked for.

When I contacted them in Bahrain, they admitted they had picked up rather random things and sent mainly some smaller stuff. And yes, the shoji screen was with them in Bahrain now, and they asked, if I could simply sell it to them? No, I didn't want to *sell* anything. I really wanted it back! They also mentioned a table I was not aware I possessed — and sports equipment. Suddenly, I realized that they had taken HUNTER'S stuff! Months later, when Hunter complained that his memory didn't work so well anymore because he could not find his sports equipment, I did not have the heart to tell him, "Hunter — it is in BAHRAIN!"

While I was working at the U of C, I was still collaborating with Professor Hinze in Basel from far distance as well. He was not my immediate PI and, being quite misanthropic, he avoided taking on any students anyway. However, a conference in our field was coming up in Mexico, and since I was then in Chicago, I could easily fly over and participate. The conference lasted about a week, and during that time, many participants got quite sick. Since they could no longer give their presentations, Professor Hinze had to do so for them. He was not delighted by that. He complained about how stupid those people were, probably not paying enough attention to the drinking water, which can cause gastrointestinal problems.

I fortunately felt fine - until the last day. My four-star hotel had a door opening to its gardens. The door did not fit snugly to the ground, so there was a gap in between. This meant an open path for any critters to enter, and when I woke up the next morning, on our last day, my shoulder hurt terribly. In the mirror, I could see a huge bump on it. I dressed and got ready to depart, joining a van ride with Professor Hinze to the capital. While he was still complaining about the stupidity of people getting sick in Mexico, I didn't feel so well myself either. I managed to get to Mexico City without him noticing anything, said good bye and went straight to my hotel room, where I went right to bed, too sick to do any sightseeing. My flight back to Chicago was scheduled for the following day. The next day, I pulled myself together while fighting several symptoms. Bad shoulder pain, dizziness, slight nausea, and severe weakness dominated the trip; and I was afraid that the airline would reject me, but no one noticed my condition. Immigration was easy and even a little bit amusing; the people in front of me, all

Mexicans, didn't know what the immigration officer meant by "visa." As a result, they showed him their Visa credit card. I made it out of the airport, where Vilma, one of Hunter's students and good friend of mine, picked me up.

When I arrived at the lab, and with the pressure of having to function taken away, I completely collapsed. Hunter himself carried me upstairs to the sixth floor, where he kept an old sofa bed at that time. It turned out that I was fine whenever I laid down, but I could not get up. Hunter became very worried and arranged to have one of his lab members, Betty, to drive me home, while he looked for a doctor. Hunter and Betty took me under their arms as I was not able to walk alone and Betty kindly took me home, which was quite an adventure because my body didn't seem willing to obey, my knees didn't seem to exist. I also carried a giant ice box with me for later. Betty's car was equipped with automatic safety belts, which even played a melody while they slid into place. The ice box caused the seatbelt to tangle and it became time consuming to get the seatbelt to function again as they played the melody over and over again but we could not get started. The scene was something of a comedy! As soon as we got to my place, I went to lay down on a sofa, which I could not leave anymore. Being nauseated by now, I did not care for food, while Betty had a good time in my kitchen. I told her to take anything she liked and just to leave me out of sight. That afternoon, Hunter arrived with a doctor. I was half embarrassed to answer the doctor's questions in front of Hunter. Finally, the doctor diagnosed me with being stung by a poisonous scorpion. Since we did not know the species and too much time had passed, the doctor could do nothing more than treat my symptoms. For the next few days, I could only crawl to the bathroom and right back. My heart didn't allow any upright movements, and I had to be careful. Hunter came repeatedly to check on me. After several days, I was slowly able to return to the lab.

Hunter could be also quite controlling. One night, Naomi from Bill's lab asked if I would like to see a movie with her on campus in Ida Noyes Hall. Naomi worked in Bill's lab as technician, but she was studying film to become a movie director. It sounded wonderful to go with a professional. When I told Hunter, he was quite upset. "WHAT!? To a movie??" I still went, but he was furious. I explained to him that I went with Naomi, but it didn't help. He wouldn't talk to me again that night. On another occasion, years later, when Joscha went through a very difficult time, I suggested the two of us would walk home together to talk. Joscha immediately said: "I don't know if this is OK with Hunter!" Hunter was the emperor — the Chicago Zeus, it seemed — and we all needed his permissions.

Competition in science is really intense. Professors must either *publish* or *perish*. Hunter said that he realized that right away when he joined the U of C and therefore always sought to get published. Other people, like Harper, an assistant professor doing microscopy, worked for seven years without publishing much. And despite his long list of accomplishments, only publications count, so he never got the recognition he should have gotten. Hunter said that he himself was lucky because Russell, now a professor at Yale, helped him a lot, while Harper had not the luck of having such great friends.

The day came when I had to fly back to Europe once again. Hunter had nicknames for every country. Switzerland was the "super tidy place" and he called Germany "Krautland." Hunter would usually offer me a ride to O'Hare airport, and it seemed natural that I would rely on him to drive me. But this time, when I knocked on his door, he wanted to have nothing to do with it anymore. "Why should I even support that you're leaving?", he asked. "I don't want to hear about Super-tidy-place, Krautland or whatever it is!", he replied. I was devastated. If I missed my flight, I would be in trouble, since my visa was about to expire. And it was never a good idea to have an overstay in the US. I begged Hunter. I told him about the consequences, but he would not budge. What followed was quite a Shakespearean drama on the fifth floor. While I looked desperately for anybody to drive me, he tried to prevent it! We could have gotten best actor awards for this. In the end, he did drive me, still grouchy and not ready to let me go. On one occasion, when we headed to the airport, he had even made me a sandwich: cinnamon raisin with onion cream cheese. I said that the sandwich really tasted funny, and he figured out that he had messed it up. At the airport, he kissed me by way of apology on my cheek. It was never easy to leave him even for a limited time.

However, Hunter had many dates. One morning, I walked into the lab to find Giorgia sitting on his lap. Giorgia was, as mentioned before, a postdoc from William's lab, where her husband worked as well. Or Hunter was sometimes happily walking home with a Hungarian lady, Joscha knew, he would even chat once with the graduate student in Lynn's lab and tell me laughingly about afterwards and on another occasion, Hunter seemed to care about no one more on earth than Gina, a tall Italian girl with brown curly hair who worked for him as a new technician. Hunter had always at least cared about my work, but it now seemed not to matter to him anymore. He would still go to dinner with me — but talked about Gina, which made me mad as Joscha jokingly once said, "Love and hate lay closely next to each other." When Matthew, Hunter's graduate student, finished his PhD and left, Gina went with him. It turned out they had been together for a while. Now the turn was on Hunter to be mad.

And of course, Hunter had several dates before. One of them was Dolores from Spain, a former postdoc in Sam's lab and now a professor herself in Detroit. She came from a wealthy family and was a triplet with two brothers. Dolores seemed the perfect match for Hunter, who kept saying he was looking for "someone smart and rich." However, rumor had it that they had been fighting all the time when Dolores was still in Chicago. I pictured her as a fierce young lady and was actually afraid to meet her when Hunter announced she would come to visit. She came in on a Saturday when no one else was around, and I was actually very surprised to meet this rather shy, skinny lady peeking around the corner. Though they had stopped actively dating, Hunter got upset when he learned that she was engaged. Her fiancé was a lawyer, and they invited Hunter to the wedding. Dolores came from a Catholic family, while Hunter never believed in religion. At the wedding, while talking to Dolores's mother, Hunter said something irreligious about the Pope, which made Dolores's mother so upset that she almost fainted. People were shaking their heads.

Hunter could also be very charming, and he

very much enjoyed the company of ladies. One time, when we all walked over to a seminar in the Knapp building, he was surrounded by all of the lab's females. He immediately commented on this, being surrounded by all "his satellites," while he was "the center of the system." We all had to laugh.

Hunter and I also spent a lot of time talking — with me more listening than talking in the beginning. We talked a lot about his past, his struggles with life's hardship, and people. I still have a book he once gave me, Colin Wilson's *The Outsider*, which he had bought in his favorite bookstore, Powell's on 57th Street. Hunter had a hard life. And honoring our deep friendship he once said to me, "You are the sister I never had."

Meanwhile, I met a nice guy. This was like a little wonder, as my many friends in Chicago made any approaches so difficult. I felt like the most protected person on the planet. If someone showed any interest in me and even thought to take me out, my friends found excuses to bring him down. I got to hear comments like: "Just a postdoc? A professor? A director? He isn't good enough for you!" With the exception of Hunter, no one was allowed to take me out.

It happened on a Sunday. I was staying at Vilma's place in downtown Chicago, where she had moved to a lovely apartment. Because I was running late, I decided to take the Red Line. I usually preferred taking the bus for safety reasons, but the Red Line worked fine, too, during the day. I exited at Garfield, where I would connect to the bus on 55th Street. In front of me on the escalator was a guy with a nice backpack. Funnily, I didn't care so much about the guy, but I rather admired his nicely colored backpack, as colors always played an important role in my life. Then he turned and asked me how to get to the U of C. He said he was going to a Native American festival being held on campus. I had not heard of it, but I simply told him to follow me, as I would take the same bus. I showed him how to pay, and we got off together. Now, where would this festival be? Though I knew Hyde Park pretty well, I had no idea. I asked people, and we finally found the place. It was indeed quite impressive. By then, I was late enough that getting to the lab no longer mattered, so I spent a little time with the guy. His name was Dan, and he was a pilot from Israel. When we parted ways, he asked for my email address, which I gave to him before returning to the lab. That was all, I thought and I focused on the lab's experiments. While Dan returned to Israel he sent me an email. He said he would not have the schedule for next month yet, his next flight would be to Switzerland, in two days. Where and when would he see me again? What he did not know was that I was also flying to Switzerland — actually the very next day. It sounded very unlikely, and therefore I had not mentioned anything when we met. So, two days later, we met again. And it was even his birthday. From then on, we got together, meeting wherever in Northern America or sometimes in Europe we could. The next time he came to Chicago, I told my friends and introduced him. Yanhong immediately picked on him, asking me if a pilot would be "really good enough?" She and Liudmila from Bill's lab came, as did Charles, a former professor in neurology. We went to eat at the Hutchinson Commons dining room, a big hall like a library with paintings at the walls, which lends the place dignity. Hunter refused to join and disappeared from the lab for the rest of the day.

When Dan was still a first officer, he had less responsibility, and we had more time for ourselves once he landed. Though he was very tired after the long-distance flight to Chicago and change of time zones, Dan insisted on going out. Once, he got tickets for a drum concert at the Chicago Symphony Orchestra Hall downtown. Our seats, purchased at the last minute and quite expensive, were on the balcony high above the orchestra. As the only ones sitting there, it was like being presented to everybody as all other visitors would face the orchestra with us sitting above. As soon as the concert started, Dan seemed about falling asleep. Now, sitting high up and just above the orchestra, I struggled to hold him back, so at least he wouldn't fall down the balcony. In the intermission, students from the lab approached us, introducing their friends or parents. And, though I tried to convince Dan that we should leave since he was too exhausted, he insisted on staying. It was the most demanding concert I ever attended!

I stayed with Dan at his downtown hotel during his visit. Naomi had kindly lent her car to us, so we could go on excursions. I still did not have much driving experience due to lack of time, and when Dan, just before he left, parked the car in front of the hotel, it was clear that I had to drive it back to Hyde Park. Driving in the US is considered not a big deal. Dan quickly showed me the functions of the car and hugged me goodbye. I drove back on the surface streets, and everything went surprisingly well. Of course — this was Chicago.

Then Hunter met Evelyn. He introduced us very briefly on the hallway of the fifth floor, so I didn't really see much of her. Shortly after, they started talking about buying a house. Evelyn, a petite woman of Puerto Rican and Polish decent, worked as a school teacher and she ran the school library. As mentioned I met her again years later at Hunter's mom's funeral. She and Hunter got married and bought a house in Evanston, a good ninety minutes north from Hyde Park on the Red Line, where I visited them in the coldest Chicago winter of 2010 and later.

BILL

Bill started at the U of C a little earlier than Hunter — around 1979/1980, as did William. Bill was Hunter's colleague and the chair of the department at some point. His lab was next to Hunter's on the fifth floor. Unlike most of us, he was born in the US. He came from Texas, and had a degree in mechanics. He was a pleasant, quiet character and easy to talk to. I had always wanted to learn calcium imaging, which his lab performed.

Bill received his PhD in calcium signaling and got a faculty position very young, again similar to William, who became a full professor at the age of thirty. Bill had two sons; because he was focused on his career, he left the children to his wife. Problems arose, they filed for divorce, and she went on to get a PhD in humanities, while the children stayed with Bill. Sadly, she developed cancer. She returned to Texas to her parents and had to undergo surgery. The cancer recurred, and Bill went back with his children to say their final good-byes. When Bill returned to Chicago, it was a hard time. He had to finish a grant and couldn't quite focus on it. So his best friend Sam stepped in and completed the grant for him. Later, Bill met Alison, and they got married. Alison worked in pediatrics administration. And they bought a beautiful big house.

One day, as one of my birthdays approached, I started thinking about what to do that day. When I first came to the US, I would tell everybody that it was my birthday, following the Swiss custom: they should be around at a certain time, so we could have some cake together I brought in. But, as Joscha taught me, in the US it was the way around, all the others treat the person having a birthday. Therefore, unless one wanted to be celebrated at others' expense, it was wise to get away. The year after I looked into my options where to go on my birthday. As a toxicologist, I read about the Kentucky Reptile Zoo, where they produced venoms. So I decided to drop the director of the Kentucky Reptile Zoo, Andy, an email, asking about visiting in early December; and, not knowing it was my birthday, he suggested exactly that day to come. Great! Now I had a plan. I flew to Cincinnati, where Andy's assistant, Heather, picked me up. It was a very nice day. Andy showed me all his many snakes and taught me how to milk rattlesnakes, how to tube feed a baby coral snake, how to catch mambas, and much more. When Andy had told me I wouldn't have to worry about a hotel, I thought he meant that plenty of rooms were available in town. But there were no hotels around. However, when night fell, he simply suggested sharing the common bedroom with everybody else, the coworkers and him. The alternative was to sleep in his office next to the twenty-three rattle snakes, and much to his amusement, I chose the snakes. It was a weird night: The snakes rattled upon each turn I made, but knowing them kept safely, it didn't matter to me. And while I slept quite well, everybody downstairs expected me to return any moment. When Andy learned it was my birthday, he gave me a good number of freeze-dried snake venoms as a gift. It was the most valuable gift a toxicologist could ever get worth a couple of thousand dollars and the potential to discover something new. Wow, what an AMAZING birthday present! Back in Chicago, it was clear that we had to try the venoms on tissue cultures; since my best friends were in Bill's lab, I decided to do calcium imaging using the venoms. From Diego in Mexico I also had a few scorpion venoms, which we included in the tests. Because I still worked for Hunter as well, I felt like Goldoni's *Servant of Two Masters* at times. It all worked out well: Yanhong and I published a nice article in JBC that was widely distributed.

Calcium imaging requires sitting in a dark room for a long time. While I ran experiments one Saturday morning, a fire alarm went off when no one else was working on the fifth floor. I carefully checked for any sign of smoke, but I didn't sense anything. Because I really wanted to complete the experiments, I went on working. Later, I heard that they had evacuated the whole building. I was probably the only one who kept working.

Bill was really like a prototypical American professor — knowledgeable, good natured, and relaxed. Coming from Texas, he was proud of having faced more than one rattlesnake in nature, and he would invite Liudmila and Yanhong to come along to a good baseball game. We foreigners often do not know much about baseball. As Yanhong would say, the rules were not important, the purpose was to have a good time! Bill would buy Alison and Liudmila a beer, pass peanuts to Yanhong, wear fancy sunglasses and shorts, talk to people in the stands, ask who is playing and where the players came from, and lead the cheers from time to time, even without any sign of scoring.

He was also known as the master of entertainment. Every New Year's Eve, he invited members of his lab and a number of other labs and neighbors as guests to a party. When the millennium arrived, I was invited to the party along with Yanhong and Joscha. That year's party theme, "ladies in red," required that we ladies wear as many red items as we could find. Since I didn't have many reds — that is to say, none at all — it caused me a headache, but Yanhong had some as red is considered a lucky color among Chinese. We managed well, lending red clothes also from friends. Before we drove to

Bill's place, which was off campus in Indiana, we quickly went grocery shopping at the COOP on 55th Street, so we would have food at home the next day. And I bought a little strawberry cheesecake for after the party; at Yanhong's suggestion, I put it in Bill's fridge for safekeeping instead of leaving it in the car. We all had a good time. We split into teams to play ping pong, at which Yanhong was a master. Later, we played several other party games, when Alison, Bill's wife, to my huge surprise, started to serve … MY CAKE! Of course, she did not know it was mine, or that I had put it there only temporarily, but from then on, Yanhong teased me every so often about cake coming up!

Bill also had a reputation as "king of BBQ," and he proved it when my friend Scarlett came to work for him. Scarlett was a student from Britain and, like me when I first came to Chicago, was hoping for a summer work experience in the US. I told her to talk to Bill. Poor Scarlett had a hard time with immigration because she honestly answered that she didn't yet know all her plans. But she made it through and started working for Bill. By this time, I was back in Europe, but that summer I visited Chicago for a few days with Debra and we passed through during a three-month driving tour of the national parks. Debra, working for the US state Department overseas, came up with this idea when she had three months of home leave. I took a three-months leave from work and we flew to NY city to start our road trip in upstate NY, where Debra's mum lived. When we reached Chicago, Bill invited all of us to his house and set up the nicest BBQ. From heart, I would have preferred to stay in Chicago, but we continued on our planned trip to see the beautiful US national parks for three months.

Scarlett worked with Liudmila, and was very happy about it. Once, when Scarlett was sterilizing slides with ethanol and fire, the beaker of ethanol caught fire. Liudmila solved the problem by simply putting a plastic cover over the beaker. Luckily, the plastic didn't catch fire, too.

After Debra we later got another visitor from the US State Department, which was Nelson. He was also a friend of mine, and though he did not work in science, his mind was all about science. I gave him a lab tour, showing him tissue cultures, calcium imaging, the confocal microscope and others. Bill also talked with Nelson for a while, and surprised about more visitors from the government he was joking that the fifth floor should buy a red carpet.

Bill was also into photography, taking quite professional pictures and giving them to lab members as gifts. He shared this hobby with his wife Alison. They even had a website with beautiful pictures which they frequently updated. As mentioned earlier, Bill had two sons. They lived in Chicago and New York City. Alison had two sons as well, one of whom lived in Hawaii. Two of their four children were married, and Bill very much enjoyed visiting his family.

Bill also knew and appreciated food. For our Friday seminars, he always ordered Thai food or Chicago pizza, which is known to be particular tasty. Often the person presenting at the seminar got less attention than the food, everyone was having a great time with big slices of Chicago pizza, cheese all dropping and Yanhong kept making fun of it.

When I was doing calcium imaging, Bill once asked me if two students could visit and watch me doing it. The next day he came with the two students. What I had done so far by routine day after day without even thinking about it, I had now to explain in detail to the students, slowly step by step – and it completely threw me off. Bill must have thought by himself that it was strange I couldn't show the students what I was presumably doing since months…

Bill was one of the few people around, who had been born in the US. As Yanhong, Joscha, and I were all foreigners, we once asked Bill, as "the real American," how to say a certain sentence correctly. Bill, probably surprised by our question and not sure what we wanted, simply said: "I don't know!"

At one point, Bill had two students, Pjotr and Chang, working for him. Pjotr, from Poland, was very extroverted. Chang was from China and rather quiet. Because I was again moving at this time, I decided to deposit my precious violin for a while in our locker inside the lab. Since only lab members had keys to this locker, it seemed like a safe place. One morning though, when I came to the lab, my violin was gone! My heart sank — not so much because of its market value, but because it was the only item I had left from my grandfather, so it had incredible personal value to me. I could imagine my violin being sold in a flea market in the not-so-safe neighborhoods nearby, and I seriously considered going out to look for it. But Yanhong had a better idea and simply asked everybody in the lab about it. And it turned out that my violin was safe: Chang confessed that, to impress his girlfriend, he had taken it home to have some pictures taken with it, which he wanted to send her!

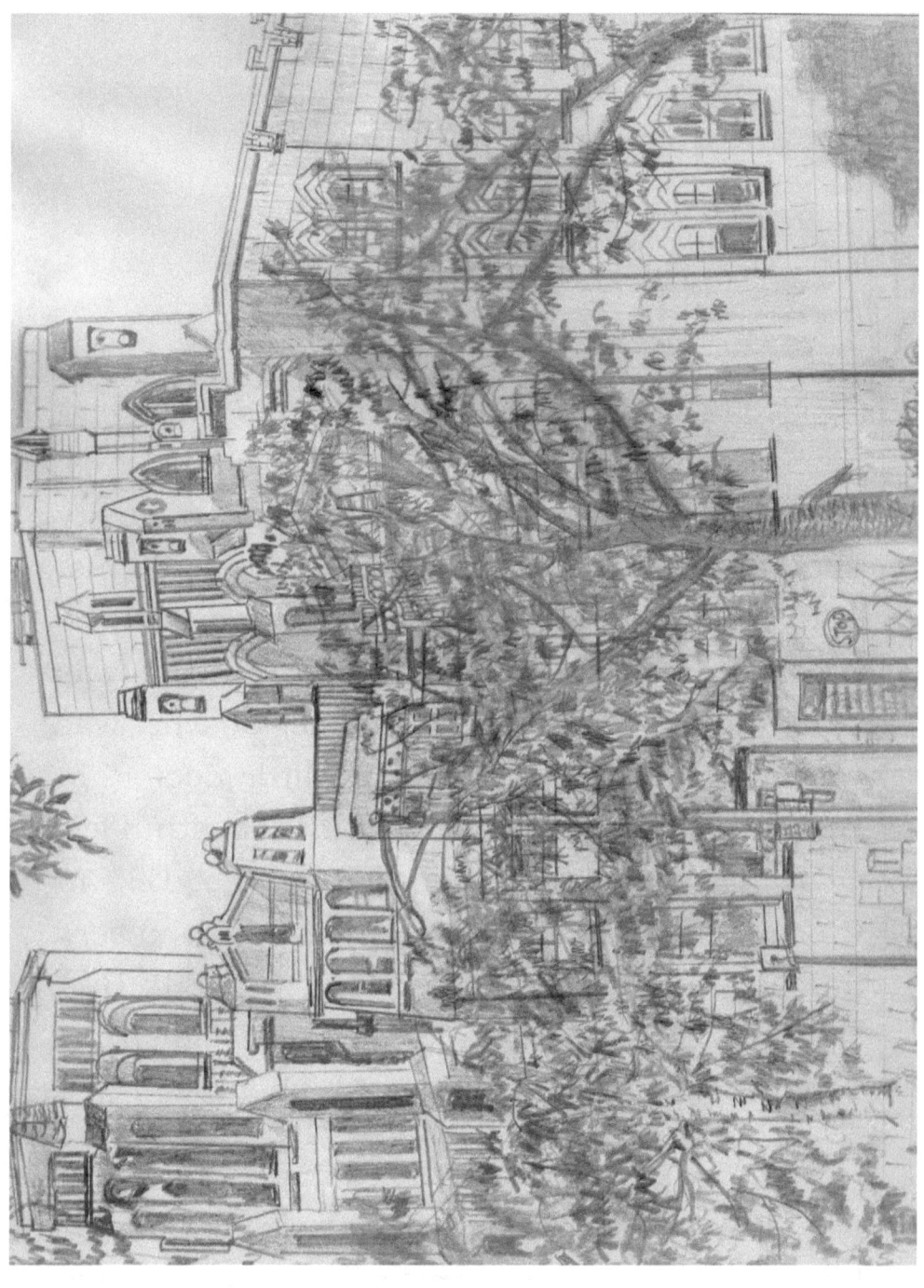

JOSCHA

Joscha came from Hungary and was very much liked by everyone. With bright blue eyes, a great sense of humor, and a touch of sensitivity and shyness, he was willing and able to do just everything asked of him. How he had come to Chicago was quite a story. While Hungary was still under a communist regime, Joscha was working on what they call the "little doctor" in Budapest, PhD students perform just within few years. He was working on a topic that Bill, in Chicago, was also working on, and he had very much wanted to exchange ideas with Bill. But, since Joscha couldn't speak much English, he could not call Bill to talk on the phone.

Then, by chance, Joscha tried the lottery; and in fact, he won just enough to buy a ticket to Chicago. So he came to Chicago to meet Bill in person. But, due to the language barrier, he couldn't say much. "Joscha!" he said, pointing to himself. "Calcium phosphatase." That was the topic which interested him and Bill was working in, too. And he kept looking at Bill. "What do you want?" asked Bill. After a pause, Bill continued, "A job?" "Hm, job," thought Joscha and said, "Yes! Job!!!"

It happened that one of Bill's technicians, Seo-yun, had just returned home to Korea, and that position became available. Now Joscha needed a work permit. He took all money they had from giving up their place in Hungary, brought his wife and son Dominik over from Hungary, and they asked for political asylum. That way, he got a work permit and started for Bill as a technician.

After Joscha learned some English, Bill asked if he wanted to become a graduate student. "Yes!", he said, as he very much wanted to continue his studies. So Joscha started his PhD. Several years later, in 1997, when he was in his finals, he received a letter from the authorities. His request for asylum was finally denied, and he had prepare to be deported. He hurried to write his thesis, which he finished in the record time of only six weeks! Just in time, again, he won a lottery — this time for a green card, which meant he and his wife and Dominik could stay.

Hunter, who was on Joscha's dissertation committee, said Joscha always gave the feeling that he was truthful. One could ask Joscha anything in science. He would either know it or he would find out, but he had not much time due to family commitments. At the end of Joscha's thesis, Hunter found a couple of things that he thought were rather odd, and he started to wonder if any proof supported them. But again, as Hunter put it, Joscha could sell anything with his easygoing way, which suited him so well and kept most people from perceiving his nervousness. And everyone bought it from him.

Joscha's father had been a judge back in Hungary. During political unrest, Joscha's father disappeared and his parents lost each other. Long after that, his mother married again. But Joscha's father wasn't dead. When they finally met again, his mother was remarried. From this second marriage, Joscha had a half-brother who ran a taxi company not far from Frankfurt. But, sadly, a family tragedy on his half brother's side broke the contact between them.

When I came to Hunter's lab, Joscha and I sometimes worked side by side in the tissue culture room next to Bill's office. This tiny room had two hoods crammed in next to each other. A person culturing tissue cannot talk, due to the risk of contaminating the tissue cultures. Hunter would sometimes stand right behind me or at the window and talk from a distance. Hunter didn't recognize any taboo, at all, and I would glance at Joscha to see if he had heard what I had just heard. Of course, Hunter was addressing me, with Joscha as an involuntary witness turning colors. Occasionally we would exchange a look, but Joscha was way too polite und shy to comment on anything.

His little son Dominik would more or less grow up in the lab. Because we all spent most of our time in the lab, it felt like we were living there, so we got to know each other very well. Joscha routinely picked Dominik up from school, and Dominik would do his homework among us. Little Dominik learned very early how to pour gels and to run them; he also learned to prepare chemical solutions and perform other tasks around the lab. Yanhong would always talk to him, and whenever I had to go to another lab on an errand, I asked Dominik to come with me. We made it a game that I would introduce him as my colleague, and everybody was very kind to him. Dominik always wished to have siblings; he might have felt a little lonely as the only child in the lab, despite our efforts to entertain him. Of course, he was not allowed in all areas, and he stayed mainly in the office to do his homework for school.

One time, Joscha and Dominik helped me transport some furniture that I had bought downtown, including the Chinese garden furniture. Dominik was as handy as Joscha and helped me to pick the right one. On another occasion, Dominik had a school project to construct a model of a house. All of us gave him some advice in passing. To my surprise, I ended up having my own room in his house.

Joscha was an excellent Dad. He completely managed his household and on top of it did everything in the lab. He was also very good with children in general and participated in reading classes at Dominik's school. One time, when Joscha had finished reading a story to the children, a little Chinese girl asked, "And what's the moral?" "There's no moral", said Joscha, not really knowing what the word "moral" meant. But the little girl persisted. "Moral, moral," said Joscha, "there really isn't!" Sometimes it is not easy for us foreigners.

None of us had much money, so we always welcomed the opportunity to attend seminars, since they always provided free food. Yanhong, Joscha, and I would carefully watch the seminar list, and we always went together, sometimes no matter what the topic – as long as nice food was provided for free! One time, I invited both of them to my place on 55th Street and Blackstone. Coming from Switzerland, I planned to have a cheese fondue, which meant cheese melting in a pot over a fire. The difficulties started with the flammable liquid one needs for the fire. All I could find at a hardware store was a five-liter container of kerosene, used for airplanes. I thought "better than nothing" and bought it. The next quandary involved the cheese. In the US, they sell something called "Swiss cheese," but Switzerland actually has many different kinds of cheese, not just one "Swiss cheese." I tried my best and bought something. Well, it was surely not my best cheese fondue, but it was by far the smokiest because burning the kerosene created lots of smoke, which in turn activated the smoke detector on the ceiling. So Joscha, the tallest of us, had to climb up and stop the smoke detector alarm a couple of times, so it became quite an active dinner. This experience was enough to make me never ever try cheese fondue in the US again. But of course, we had a great time full of laughter.

Another time, I brought in a small cake I was going to eat for lunch. I placed the cake on the table in the lab that we used for our lunches. At lunch, I ate it. When Joscha and his Hungarian friend Tamás stepped by later, they asked me where the cake went. I told them that I ate it for lunch, and they both looked at me in disbelief and called out: "The WHOLE CAKE??" I felt a little embarrassed I had not offered them anything of it.

Joscha was also very athletic. He showed me once how to juggle using huge two-liter measuring cylinders and just at that moment Bill came in, looking as if he wouldn't believe his eyes. Joscha frequently took Dominik camping, skiing, and canoeing, and he showed me how to skateboard. While he would elegantly ride the skateboard, I would go in funny lines. However, I bought a pair of rollerblades not long after and tried them over the weekend, when fewer people were walking the streets of Hyde Park. I quickly managed the smooth sidewalks, but when it came to uneven parts, especially leaving the sidewalks to cross streets, I must have provided observers with a funny picture. I remember one time that I scared a big Chinese family out of my way before I became more professional in it.

One Sunday, when the lab was in the Knapp building, I decided to rollerblade to the lab to take care of a tissue culture. The camera at the building's entrance caught and recorded me, and unbeknownst to me, Joscha, who had skateboarded in shortly before I arrived. The next morning, we were called in and lectured about how we scientists should not enter on roller blades, skateboards, etc. Anyway, I'm sure Joscha looked more professional on the recording than I did!

Joscha could also speak German, and he played piano beautifully. I once heard him playing Tchaikovsky on a key board and urged him to continue with piano studies. It seemed like anything Joscha did, he did perfectly.

Then Joscha's marriage came to an end. Joscha went through a very hard time, which none of us had ever wished for him. We would often walk home together during that time. Once, he asked me something private, and I helped him to find out. But things got better for Joscha again; life was meant to be good for him. He soon met Zorica, a beautiful girl from Serbia, who lived right across from his uncle outside of Chicago. Zorica and her sister lived with their mother. Joscha married Zorica, and Dominik's wish to have siblings came true: they had two children together, Lilly and Adam. We always joked that the next girl would be named Yanhong.

Joscha and I had one trait in common; we both

ate very slowly. Once, when Yanhong and I visited Joscha at his home, his wife Zorica ordered pizza. After everybody was already done, Zorica jokingly commented that Joscha was still eating, as was I, without having eaten much so far.

Both of us also had a great interest in and passion for chemistry. As students in high school, we both participated in Europe's Chemistry Olympic games — and we had probably even met before in the mid eighties in one of those Olympics.

When Joscha finished his PhD and left the U of C in 2000, Yanhong and I wanted to give him a nice present. Yanhong had a great idea: to get him a blank book in which to put wishes, sayings, and comments from all of his colleagues. We bought a nice leather-bound book downtown. Then we decided that I should go around and collect everybody's contributions. Oh my, I found this very hard because I was rather shy! But, for our best friend Joscha, I would do it. Of course, he had no idea, as we planned this as a surprise. And sometimes, when someone would stop by at Bill's lab to write a contribution and Joscha would also happen stepping by, we ended up standing there as if we had nothing to do in front of him. When we had collected everyone's saying, Yanhong and I filled the book up with some quotations from science and medicine. Our favorite one was: "In Medicine: Before thirty, men look after disease. After thirty, disease looks after men."

Joscha had been applying for several jobs, among them a position at Harvard Medical School. He even got the position while still working on his PhD, but he didn't accept it because his former wife didn't want to move. I had been working at Harvard Medical School before and left to be back to Chicago. Now Joscha had his chance and stayed in Chicago. Chicago was simply our grounds! So he sent an application to Argonne National labs. They asked him to suggest a salary, and having nothing to lose, he asked for more than he would expect. To his surprise, he was accepted right away, and he took the job at Argonne. That meant, we had to travel a bit in future to see Joscha off campus.

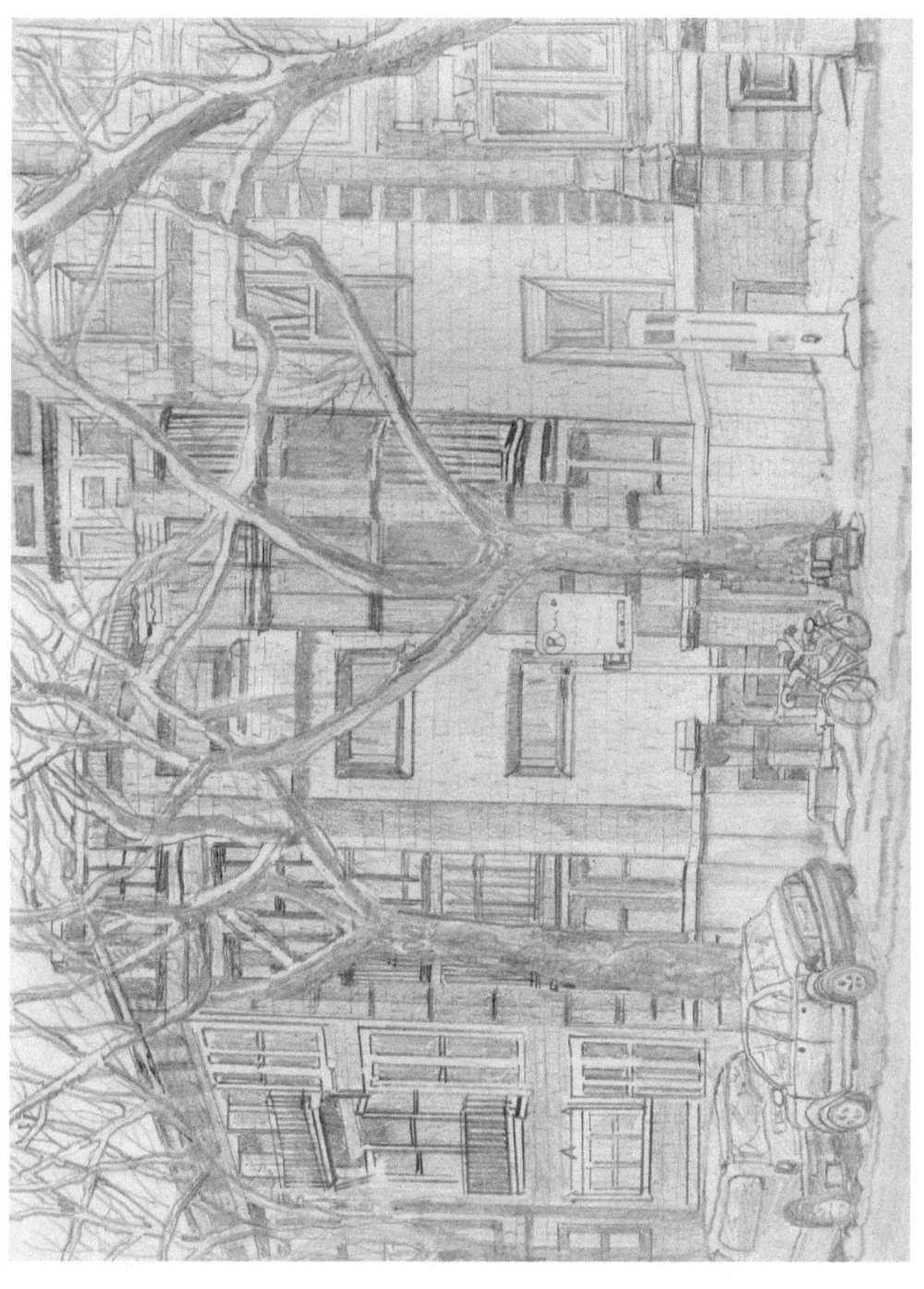

YANHONG

Yanhong was the friend I always stayed with when I came back to Chicago for shorter visits after I had taken on a position outside the US. Yanhong came to work for Bill in 1997 as a research associate, which was probably one of the greatest gifts ever received by Bill, as Yanhong was very ambitious and professional. She would lecture me at times that I should focus more on science or medicine alone, because I always preferred to be a generalist rather than a specialist. But since I didn't want to pursue the same path as her and become a professor, it didn't matter.

We met when she arrived on the fifth floor in July of 1997. She needed some paperwork done for registration. Since I knew where to go, I offered to walk with her to get it done and we talked all way long.

Yanhong came from a big family in China. Her grandfather on her mother's side graduated from Columbia University in the 1940's and actively pursued social work — working at hospital organization and for charities. He must have been very social as well, because he stopped by Einstein's house in Princeton and just asked to have their picture taken together - and Einstein even agreed! Sadly, the picture got lost over the years. Her mother's family was well educated. And her father's family produced many kinds of leaders. Her father, at the young age of thirty-two, became the head of the Institute of Mining Research. This area was very important in China at that time. Yanhong learned that a leader is a person who everybody wants to work for.

Yanhong was young, living with her parents and two siblings, when the China's Cultural Revolution started. All schools were closed for ten years, and the children had to study at home. Nobody could find books to buy, but luckily Yanhong's uncle worked at Beijing University and could still provide the family with books. Of course, these were not books appropriate for children; in the third grade, Yanhong was learning what children learn in middle school today. During the Cultural Revolution, all teenagers were required to spend two years in the countryside. After that, they could take on any job they wanted. When the day came for Yanhong to leave home, the atmosphere was similar to the 4th of July in the States, with a lot of people on the streets and many excited children, but with several parents crying. Yanhong's father accompanied her to the truck which would take her to the countryside. Her mother, however, couldn't come because she was crying so much. Her father told her two things to remember: First, money was never a problem, so she didn't have to be scared. Second, she had to come home every month. That way, the family would not become estranged from each other, and her parents could keep an eye on Yanhong's development.

The ride took all day, from morning to evening. It started with a truck ride to the Yangtze river, followed by a ferry crossing and two buses afterwards. The children finally arrived in a village in a mountainous region and were directed in the dark to a simple place to eat. Yanhong could see how poor the people in the village were;

they had neither electricity nor running water. Along with six other girls, Yanhong was sent to a Youth House, a two-bedroom house with small windows. Lacking electricity, the girls had to use candles. They had a stove, which worked simply by burning dried grass, and the girls had to wear long sleeves to protect themselves from the mosquitos. The girls didn't know each other, and facing these poor conditions, all of them except Yanhong started to cry. Yanhong told herself that crying wouldn't help; they couldn't go back. Instead, she thought about how to make life more acceptable.

Meanwhile, all of the village people came to see the newly arrived girls. The people were very simple, without requirements nor expectations and warm-hearted. They had never left their village and couldn't read or write. They invited the girls to their houses for dinner. The next morning, Yanhong got up early and went to the fields with them. She noticed the freshness of the air and the beauty of the sunrise and the singing birds; life wasn't so bad after all, one must always see the positive side. As the village had a shortage of girls, all the boys were happy to help them. That first morning, Yanhong learned to get water from the ponds and how to carry it home on a wooden holder over her shoulders, enabling her to carry up to 120 kg. It took her an hour to learn. It wasn't easy, but then she was very happy she learned it. She also learned to brush her teeth at the ponds. The village had ponds for drinking water and ponds for washing, and while Yanhong made a lot of progress at learning all of these new things, the other girls were still asleep in the early morning.

In their first meeting later that morning, Yanhong suggested that the girls should consider these two years as an exercise that would enable them to live in any condition afterwards. And she repeated that the government had promised them that they could get any position they wanted after the two years. Yanhong in memory of her grandfather volunteered to act as their leader for the first month, suggesting they take turns afterwards.

Their daily routine was to wake up at 4 am and then to prepare rice soup for breakfast. Still today, Yanhong never fails to eat rice soup in

The Fifth Floor

the restaurants in Chinatown. She continued to be the only one who didn't cry and who could function. All the other girls were too paralyzed from sadness, but they developed a routine: one of them would stay home, get water, and cook, while the others worked in the fields.

Soon, Yanhong became ambitious and wanted the girls to raise their own chickens. One guy helped them to make a cage for many chickens, which produced five to six eggs every day. Yanhong even got them their own pig, geese, and ducks. Later, to do more to support themselves, they created their own garden, where they grew spinach, vegetables called "Chinese greens", tomatoes, long green beans, chives, cucumbers, squash, sweet potatoes, peanuts, and sunflowers. They sent the sunflower seeds to a factory that produced oil. Over time, their garden became quite big. They killed the pig and produced salted pork.

Yanhong soon met with the leader of the village. She was given great opportunities to give speeches and lectures in some counties of the state. Because of her excellent performance,

Yanhong was selected as Director of Women for her county.

The girls thrived during the first year mainly because Yanhong's guidance was outstanding; no one would cry anymore. When the girls went home for visits, each one would proudly carry five pounds of vegetables that they had grown to present to their parents.

While the villagers taught the girls farming, Yanhong taught them how to read and write. They set up a school in their house. In the winter, classes took place inside the house; in summer, outside. Soon, they started to teach the village boys. Later, everybody came to class. Yanhong learned leadership skills, remembering her grandfather's words.

After the two years were over, the universities reopened for the first time in ten years, and thousands of students applied. Physics was very popular, along with chemistry and math; architecture, teaching, and medicine only ranked secondary at that time. Yanhong considered studying physics, but an earlier incident in her village changed her mind.

Two village boys, twins, treated the visiting girls like sisters. They would go to the girls' house and always offer their help. They would carry water for them from the pond and help them with the harvest and whatever else the girls needed. They became the girls' best friends, but then they fell sick. The next hospital was at least 200 miles away, an entire day of travel; since the village had no car, the boys suddenly died. Yanhong had been to many weddings before, but never any funerals, and she was very sad. The funeral took place in a chilly wind. As is customary in China, a pig was served for food. Whenever a life ends or a new life begins, a pig will be served in China. The parents were deeply saddened, as they had lost not just one child, but two. Because life in the countryside was so challenging, farmers would always have many children despite China's campaign of the one-child family. This incident made a big impression on Yanhong, and she decided to forget about physics and study what was really needed — medicine.

At about the same time, another incident occurred. Yanhong's mother got sick, and Yanhong had to go home. Her mother had hot flashes, tachycardia, and mood swings. She went to the hospital more often than before and wouldn't believe her doctors, who explained to her she was only experiencing what all women do at a certain age — the natural course of the aging body. In the same apartment building, a lady doctor lived from Yanhong's father's institute, who came to see Yanhong's mom whenever needed. She patiently explained to Yanhong about the hormone changes, and Yanhong was very grateful for her help. This strengthened Yanhong's decision to study medicine.

Getting into a university was not so easy at that time. First, all of the young people who had been sent to the countryside wanted to return to the cities, and entering a university provided one way to do so. As a result, all of the young people headed to exam rooms to take the entrance exam. Of course, only ones who were prepared had a realistic chance, and Yanhong had been away from school for the past two years. However, she had never forgotten her uncle's words: "A mind is like a knife: if you want to sharpen it, you have to do something, and that is to ask questions." So, whenever Yanhong had a moment for herself, she read books and asked questions, even though surviving life in the countryside took a lot of time and effort.

The closest exam room to her village was at the top of a hill, and Yanhong had to line up with a crowd of people to get an ID to register for the exams. Then the entrance exams started. They were spread over three days: a half day for mathematics, another half day for another subject, and so on. The questions were very difficult. Afterwards, Yanhong had to wait a couple of months to learn the results. Despite later taking many exams in her life, Yanhong never forgot this one. She was nervous. Her parents told her that she tried her best and that if she did not pass, she could always try again next year.

When the day came for the results to be announced, Yanhong wanted to go to the countryside to see if she passed. Since her father was worried about her, he decided to find out the results by phone and, if necessary, to go to see Yanhong as well, to make sure she was OK. The chances of being accepted were very limited, as this was the first national exam in ten years. Less than one percent of those taking the exams would gain admittance to university. The results were listed by scores rather than by names, and Yanhong started to read the list from back to front, beginning with the lowest score, hoping she would just make it. As she worked her way towards the front of the list, not seeing her name, her courage began to fade. By the time that she reached the first page, she was almost sure she had not passed. Then she suddenly saw her name — on the first page! Doubt arose in her mind; perhaps this was another person with the same name? This had actually happened to her several times in her life — even later, in Toronto, she came across another person with the same first and last name.

While she was surrounded by so many familiar faces, she didn't want to see anybody. She first wanted to make sure that it was her name on the list. It took her another two hours to find out. She had to go to the registration office, gather her courage to knock at the door, and compare the place of residence of the listed name. The woman in the office finally confirmed that she had found her own name on the list.

By now, the last ferry crossing the Yangtze had left, so Yanhong could not return home that night. Luckily, one of the seven girls from her village lived near the ferry terminal, and Yanhong stayed with her overnight. They stayed up all night, and the girl started to cry because they were losing Yanhong. For Yanhong, to stay in the countryside had not been an option; everybody in her family went to university, and sooner or later she planned to do the same. What made this situation unusual was the setting: among the thousands of people from all of the villages around who took the exam only Yanhong had passed the exam — the first exam in ten years. When Yanhong came home to her parents the next day, they welcomed her with a big dinner celebration.

After the exams, the students had to list where they wanted to go. Yanhong's only choice was medical school. When she got to talk to the official who recruited all the students at university, he told her she was very brave to give only that one option. Yanhong replied: "Well that is what I want to do. If I don't get it, I may have to wait another year." The official selected her. In the countryside, Yanhong had learned two things: first, how to survive; and second, how to manage people. Now, at university, other new things would await her.

Meanwhile, it was time to say good-bye to the village. Yanhong's mother had never been to the countryside, and Yanhong invited both of her parents to the farewell party. The Chinese custom is to kill a pig only when a life begins or ends. Now, for the first time, the village killed a pig for a farewell — Yanhong's farewell. None of the six girls she had lived with would come to the party; they were too sad about losing her. With her father's help, Yanhong packed all of her things and then went to the celebration with all the farmers. For her mother, everything was new, and she said that life in the countryside was indeed very beautiful. Yanhong returned home by truck and afterwards lost track what happened in the village as she would focus on medical school now.

Yanhong had a good time at medical school; all students in her class were eager to learn. For her, it was a dream coming true. She graduated from Tianjin Medical University in China and received her MD degree. All of her classmates received leadership positions of all kinds; all became deans, and one even became a minister of health. Her class was the first to establish a new faculty in ten years, and graduation was a big event. Eleven of her classmates stayed in the same university; all of the others went to other hospitals. Yanhong was one of the eleven to stay. She worked for four years as a junior member of the faculty. But, sadly, her father passed away from esophageal cancer during that time in the early eighties.

One day, the university president called Yanhong into his office. Her medical school had a foreign affairs office; they also had an eight-year training program granting a combined MD and Master's degree. All classes were taught in English, as foreign professors were invited to lecture. The president asked Yanhong if she wanted to become a representative of the foreign affairs office. For this, she would need to teach and to prepare tests with another colleague. In addition, they would expect her to entertain foreign professors in her spare time. Yanhong accepted.

Yanhong went to the foreign professors and asked them how she could help them. One answered that his wife was interested in Chinese cooking which she wanted to teach back in the US. So Yanhong asked her grandfather for help. He knew every single restaurant in town and could help her to choose ones with special dishes. She also arranged for recipes, making the professor's wife very happy. Another professor expressed an interest in Chinese opera. Yanhong biked to the opera and negotiated for tickets. She explained that they better sell her tickets for these foreigners, because the opera represented their country and a newspaper would report on it. Humble, she only asked for four tickets — for the university president, the department chairman, and two professors and left herself out. In addition, she obtained special permission to have pictures taken of the foreigners with one of the actors who was quite famous at the time. For the day of the performance, she arranged a car with driver to take them to the show, and she took all of the pictures of the group with the actor after the performance. The president was very pleased and decided to set up an exchange program sending Chinese students to the US.

He called Yanhong and said, "You have worked so hard for us now for almost four years, so we decided to send you." Yanhong was shocked. She had considered the position only as a job and wasn't planning to go to the US. The president said, everything would be set up by the office, and they would also provide her $800 to buy a suit. Yanhong so far had never dressed up, as people weren't used to it in China at that time. The president added that everything had to be done in one month, which came as another shock. Though Yanhong learned to speak English with the foreign professors, she never really had any training. Her grandfather simply said: "Don't miss your chance!" The first year would be paid by the exchange program, and Yanhong left in 1986, one year after her sister left for the US and one year before her brother left, too. Her mother would also join them later, early nineties. Yanhong has never returned to China again. She arrived in the US on a Monday, July 4th, landing in Louisville, Kentucky.

After her year in Louisville, Yanhong was prepared to go back to China. But her American professors suggested to get her PhD in the US instead of returning to China. Yanhong once more thought she had just done her job, and she had planned to become a professor in China. However, whatever job Yanhong took, she always tried her very best. She had only published one or two papers by that time, but she did well in teaching and had taught several classes. She excelled in the practical part of her medical program and had seen many patients, which added to her experience. As a junior faculty member, she had done administration without payment, and she had learned how to deal with people. Social skills are very important in universities; people come from many different places and the language they have in common is science. Yanhong learned almost everything, with the exception of the English language. Her English at that time was OK for China, but not for the US. Her mentor in Louisville took her aside and asked her to explain a science question for five to ten minutes.

When Yanhong couldn't, her mentor said: "You may be able to answer the question in Chinese, but your knowledge of English is not enough. Studying for a PhD will teach you to talk about any science question for a few minutes. If you are required to give an hour-long, or even a day-long presentation, of course, you have to prepare for it. But you should be able to answer a science question for five minutes. Studying for a PhD will also teach you communication skills. Answering is only one part of what you must learn, engaging in discussions is another."

So Louisville became Yanhong's first stop in the west; her professors recommended to continue on to graduate school. She took the GRE and TOEFL and applied to universities within driving distance of Louisville — in Pittsburg, Indiana, and Minnesota. And she got a three-year fellowship in nephrology, which she used for the applications. Then she went to church, not because of religion but to socialize — to meet the professors there and to see people from different departments. In addition, she studied in the evenings in the library with her roommate, Hao. While Yanhong studied for graduate school, her roommate prepared for the board exam in medicine. Her roommate, an oncologist back in China, had left her three-year-old daughter and husband behind to prepare for FLAT, which she finished in only one-and-a-half years. When her family arrived, they moved to South Carolina, and she paid a lawyer to help her get a green card while she completed her residency. She practices oncology in Cleveland today.

Yanhong's mentor wrote a letter to the university president, and she headed for her PhD. Her mentor asked which field Yanhong preferred, and she told him she particularly liked physiology and nephrology. He recommended to study with Professor Bernstein in Toronto, the only renal physiologist in the area. He was of German descent, and as it was custom in Switzerland, he was commonly addressed not by his first name but as "Professor Bernstein". His discovery of the atrial natriuretic peptide (ANP) in 1980 had made him famous, earning him all kinds of awards. So Yanhong's mentor called up Prof. Bernstein.

At that time, there was a FASEB (Federation

of American Societies for Experimental Biology) conference every April in Washington DC, and Prof. Bernstein asked Udo from his lab to interview Yanhong there. Udo kindly invited Yanhong for coffee. Five months later, on Labor Day in September, Yanhong left for Toronto.

Yanhong found life in Toronto as a graduate student very interesting but challenging. It also provided Yanhong with great opportunities to build up her network. She met Emma, who became her "contact family". Emma introduced her to her church. Yanhong accompanied the priest and a lot of friends for a short trip to the countryside of Ontario. Emma loved to travel. Every time, she returned with gifts, books, or photos for Yanhong. During their dinner conversations, Emma "transported" Yanhong to Europe, Australia, or Africa. In addition, Emma provided Yanhong with furniture.

Yanhong also met Tanner at Toronto's Department of Physiology, where he worked as a professor on the same floor as her lab. They played tennis and squash together and went sailing and to movies. As soul mates, they became very close friends and, later, more.

But, at that time, Yanhong was very popular and was dating a Chinese guy everyone called "Bad News." He got this nickname because, whenever he showed up, he took Yanhong away — "bad news" for the lab! He was a graduate student who had started at the same time. "Bad News" worked in heavy industry, and his building was not on the campus, so seeing Yanhong required planning. He always wanted to return to China and finally did. Yanhong didn't want to return; they later lost contact.

The University of Toronto attracts scholars from all over the world. While there, Yanhong met her classmates Hao and Wang. Both were postdocs at that time. Hao had received his PhD from the Netherlands; Wang, from Germany. Later, Hao became a full professor and the director of the transplant unit at University of Toronto; Wang became a director at Roche, a large pharmaceutical company.

Yanhong stayed in Toronto for six years. Because she had not studied in Canada before, she also had to earn her Master's degree, which she did in only one year, and she added one year as a postdoc in Zachary's Toronto lab to write up her results. During this time, Yanhong became a Canadian citizen. Her next goal was to get a US green card, so she could practice medicine in the US.

After her PhD, Yanhong applied to several places, including Abbott Laboratories in Industry, in their Department of Development and Research. But they sent Yanhong a letter saying that she would be overqualified, so they didn't hire her. Anyway, it would have not suited her plans. In order to get a US green card, she needed a university job. She applied to Harvard in Boston, and, in Chicago, to the University of Illinois at Chicago (UIC), Northwestern, and the U of C. She had planned to spend only one day interviewing in Chicago. She started with the interviews at a lab in UIC. Then she went to see Bill.

As in Joscha's case, a technician had just left, so Bill had an open position available. He was thinking of getting a postdoc instead of a new technician. Bill only asked for a letter from her previous boss. He offered whatever she presently had in Toronto as salary plus 10%, because living in Chicago was more expensive. Yanhong's PhD had been in integrated physiology, while Bill's lab focused on the cellular and molecular levels. In fact, Yanhong wanted to learn something new. She asked in which journals Bill's lab published, since she wanted to have good papers. Bill replied "JBC — the Journal of Biological Chemistry," which was quite a highly ranked journal.

At the same time, Ka, Yanhong's brother, got a job offer in downtown Chicago; Yanhong's sister lived north of Chicago near Evanston. So Yanhong accepted the position in Bill's lab as a research associate at the U of C. The three siblings had a big dinner celebration then in Evanston north of Chicago.

While I worked for Hunter, I always went over to Bill's lab to see Yanhong at least for lunch. She would bring in the New York Times, and we would look at outdated horoscopes, making fun that the predictions hadn't happened yet. Later, when I worked for Bill, we could simply talk more in the lab. Yanhong had many Chinese connections, and we often fantasized about going to China one day together, where one could even get laser eye surgery from one of her friends.

Yanhong's brother, Ka, still lived not far from Chicago with his wife and his son, Kason (a play on words, as they chose "Kason" because he was the "son of Ka"). Their sister had moved by now, having moved with her husband and two daughters to New Jersey. Yanhong's mom, Mainli Yang, who was an artist, lived not far from them. She did beautiful Chinese drawings and had her own exhibitions. Yanhong's sister even urged her mother to do drawings as gifts on commission, and Yanhong had albums of her art reprints. And kindly, she not only created one for me personally, but also provided all the drawings in this book.

From Yanhong I learned many things, including Chinese cooking, which I continue to do this day. We would drive up to Chinatown together for grocery shopping and restaurants; often, I was the only non-Chinese person there. One time, I ordered a special drink with algae or some kind of grass in it — which made me the center of Chinese attention, since it was a drink that no one else usually orders. And we would go for dim sum on special occasions like my birthday. It was so much fun when they came to the table with a tray full of covered dishes for us to choose from. Once, when the waiter brought her tray to our table and I found something very delicious, I wanted to order it again. Unfortunately, we had no idea what it was called. Yanhong asked everybody, including the restaurant manager, until we got it. Chinese persistence — amazing! We would also go to Chinese bakeries, which not only sell sweet goods, like the buns I always liked to get there, but also chicken feet, which I already knew from the Chinese moon festival. The most bizarre places, however, were Chinese butcher shops. Yanhong went to buy a half duck, and I cautiously stopped outside, reading signs like "spicy pig ears" or "pig nose."

One time, in a market I saw beautiful blue and red crayfish with their eyes moving up and down. I couldn't resist and purchased one and brought it to the lab the next day. Now Bill's lab had a new pet, and Yanhong would say, "Crayfish soup is coming up soon!" Well, there was no soup. It was the lab's pet. But, since we didn't know how to feed it properly and winter prevented us from freeing it in Lake Michigan, sadly, the crayfish didn't live long.

One particularly hot summer, we decided to buy an air conditioner for each of our apartments. Yanhong found some ads, and then we went to buy them. They came as huge and heavy as an engine of a Boeing, and we had to ask others to help us carry them from the car to our places. They were also noisy like engines of a Boeing, but they worked well. In the US, air conditioners just sit in the window, and I was always afraid that mine would fall down one day and kill someone on the street.

Yanhong and I had in common that we never in our life drank any alcohol. Hunter had some wild years when he was young, but this was a long time ago and Joscha told us laughingly that he had tried it once when young and ended up dancing on a table. Yanhong sometimes joked, after finishing a project, about "getting drunk," but this was really beyond our imagination. Whenever we attended a seminar downtown with others, Yanhong, who had a car, always became the designated driver.

Occasionally, we both liked to watch a movie or the ER series, shot in Chicago, on TV at home. We would also go downtown to a movie theater, where we would try to see more than one movie at a time, to make it a real "movie night". Like Hunter, Yanhong knew a lot about movies.

She also knew a lot about photography. Later, whenever I left Chicago and Hunter was living in Evanston, she would drive me to O'Hare; each time, we would spend some hours in the Morton Arboretum, not simply taking pictures, but pictures in different fashion style, we kept changing clothes in between. For more exciting sights, I even ended up climbing trees in high heels, like a goat, despite not being in my twenties anymore.

As foreigners, neither of us had any idea about Halloween in the beginning, and so it happened that once, on my way back to Yanhong's place, where I was staying at that time, two children stopped me between 54th and 55th Street on Woodlawn. "Trick or treat!," they said. I asked them what they wanted. "Trick or TREAT!!" they screamed, and though I had heard the words very clearly, I had no idea what it meant. So they patiently explained the whole procedure to me. All I could do was offer them to follow me home. When I arrived at Yanhong's place, she was quite surprised that I came with two children. "Do we have some candy?", I asked her. Now, Chinese in general don't eat sweets much, and we ended up looking in every possible drawer for candy. We finally found some old ones, and the children left a little disappointed that these people knew so little about American customs.

Another custom which was new to us was the phone system directing people by pressing buttons before reaching an operator. We always made fun of it and would say in daily life, "press one" or "press two."

As I had been always interested in chemistry, I became a member of the American Chemical Society. As an honor, I ended up being listed in the "Who is who". At that time the

Chicago FBI was looking to hire scientists and not long after I got an invitation by the FBI, who ran an evening of job information. This event was downtown and Yanhong was so kind not only to drive me, but to wait until we were done and to drive home with me! The job sounded interesting, but one has to be US citizen, which I wasn't. So I stayed working at the U of C.

Illinois requires drivers to renew their driver's licenses every four years and requires that they come in person for a vision check every eight years. When eight years had passed, I went with Yanhong to the driver's license office downtown. After the vision test, we were sent over to the cashier. "Ten dollars, please!", said the lady, a tall African-American woman. But Yanhong and I, not used to slang, didn't understand her. "What did she say?", I asked Yanhong, and she answered "I have no idea!" And we looked expectantly at the lady. "Ten dollars, please!' she repeated, but we didn't understand a word this time either. "TEN DOLLARS, PLEEAASE!!", the lady almost yelled at us, desperately, as if we were deaf. Reluctantly, I pulled out a fifty-dollar bill, just to give it a try. She was obviously relieved when we foreigners without language finally left.

Yanhong and I always went biking along Lake Michigan on weekends. We would meet as early as 6 am, always stopping for a milk shake at the Navy Pier on the way back. The only problem was that I had no bike at this time. My place on 55th Street and Blackstone had a huge basement, where everyone parked his bike. We noticed that one of the bikes there, owned by an old lady, was never used. Yanhong convinced me to "borrow" it. So we would go biking with the old lady's bike and return before she woke up and knew about. But I didn't like the idea too much, and I started to look for a used bike in a yard sale. Not long after, I found one, though it was actually a child's bike. I looked like a circus bear on this little yellow child's bike, but a bike was a bike, and even more important, it was my own! Now I worried that somebody would see me on this tiny yellow thing. Again, Yanhong convinced me that no one would be up so early. But, as things go, Joscha's Hungarian pal Tamás, who worked for William, once met us on the way back. I was pretty embarrassed. Even worse, one morning when Yanhong and I went biking, I would get even more attention. That morning we suddenly found our bike path closed off by guardrails. We were not sure what to do. There was no sign, so Yanhong suggested simply to climb over it. I followed, and we rode on. While talking, we didn't notice that we were actually the only ones there. Usually, we would see other people biking or jogging as well. Then suddenly someone passed. "On your left!" And another one. And yet another one! All looked very sporty and professional, and when we entered downtown over the little bridge, we saw all the TV cameras: we were in a race! Of course, once we got on the bridge, we couldn't go back. We had to ride in front of the cameras — and here I was, pedaling this ridiculous little yellow child's bike! The next day we got to hear teasing comments from all our colleagues who had watched the race on TV.

As we both were in medicine, sometimes, when we walked home together, we would play "diagnosing people." One of us would look at a person passing by and ask the other, in a moribund voice: "How much longer?" And the other one would come up with an estimated lifespan, including the diagnosis. Occasionally Roman, James's colleague, would turn up with a loud "hello" on our walks home, always seeming to come from nowhere and taking us by surprise, and we would just end up laughing. Or we would drive to a car wash, which none of us had really done before. Chicago presented us with many practical things we still had to learn or get used to.

When I worked in late Hunter's lab, not getting home before midnight, Yanhong would sometimes stop by and bring me a dinner that she had freshly prepared at home. Between having my own homemade sandwiches, sharing doggie bags with Hunter that he brought in from the day before, and Yanhong's wonderful Chinese food, I never had to starve.

One day, I met Sanja, a postdoc from Croatia who worked in another building in Warren's lab. Sanja and I occasionally met for coffee. Soon after, she invited me to her birthday. Sanja lived off campus, and had a nice little gathering. That was where I met Tom. Joanna and Tom had been classmates and had known each other for a long time. Tom, who had just been through a difficult relationship, worked for Siemens. He and I decided to go ice skating the following week.

I asked him to pick me up from the lab and gave him directions. On that day, I had to work upstairs on the sixth floor. I told Yanhong that I was expecting someone and asked her to send him up. When he arrived, Yanhong gave him a very hard time. She started by asking him if he had an appointment with me. Then she lectured him, saying that he couldn't just walk in like that and asking if he knew who I was, etc. All of this intimidated Tom, and, to Yanhong's amusement, he pulled out a printout of my email. Yanhong told him to wait, she would need to check if I would have time for him. Meanwhile, I had no idea that any of this was going on. When we finally met, I offered to give him a tour of the lab, and he soon noticed the danger signs at the doors warning about radioactivity. As we work there every day, it is nothing special for us, but since it was all new to Tom, I showed him how a gamma counter works. It was just too much for him — the danger in the lab, the encounter with Yanhong ... instead of going ice skating together, he ran off! And I never saw him again. Yanhong simply said, he wouldn't have been appropriate even just as a friend, and we still laugh about it today.

As we often stayed in the lab until midnight, we considered taking the U of C taxi service, which will pick up someone from a university building and take him or her home. U of C operated buses, as well, but they only run during peak hours. Having used the free taxi service a couple of times, I had an incident. After the driver had dropped off Yanhong, he suggested that he could drive me downtown to a bar. But I didn't want to go to a bar. All I wanted was to get home, so we argued while driving around Hyde Park. Meanwhile, Yanhong had already tried to call me at home to make sure everything was OK. When I didn't pick up, she started wondering. Meanwhile, I grew desperate and tried to open the door of the taxi while driving, simply to jump out, but I didn't succeed. Luckily, the driver finally gave up, and when I reported the incident next day, he was fired. From then on, we hesitated to call the service again.

Safety was always a priority in Hyde Park, and therefore the U of C campus had its own police to provide additional protection. Yanhong told us that some of the police officers worked for the Chicago Police Department and, in their free time, worked extra hours for the U of C police.

When I had to leave my apartment on 55th Street and Blackstone to finally move back to Europe, Yanhong came to help. I had to return to Switzerland for my career, if I did not want to keep renewing my medical licence as is required in the US. And of course, it was nice to see Radu, and Liesel. But it was biased. My heart was so sad having to leave Chicago that I put off packing. Three days before my flight, Yanhong took in my plants, I had carried all way from Ford City once and which had been with me since. In addition, she suggested collecting boxes at the hospital that night, and we started our project the same night. She parked at the entrance of Abbott Hall, and I ran upstairs to collect anything left in the hallways to be thrown out and bring them downstairs. Yanhong loaded them in her car, and we drove to my place. When we returned for a second load and started to drive, the police pulled up behind us. To our utter surprise, they turned on their flashing lights and asked us to stop. So we did. Neither of us had ever been stopped by the police, so we didn't know what to do next. When we started to get out, they instructed us over the speakers, "Stay in the car!" Then a police officer approached Yanhong's window. Both of us just stared at him in disbelief. "IDs, please!" he ordered. When he saw mine, which actually said "Hospital," he smiled. "Oh, all right, then," he said, and let us go. But the two of us were still in shock.

Around the turn of the millennium, Yanhong was dating a professor in Las Vegas. We immediately started to tease her about gambling coming up, but due to a sad incident the relationship ended before it got really started. Joscha and I were also having a hard time: he was getting divorced, and I had just lost my family in Switzerland.

And later, sadly, Bill lost his grant. After several failed attempts to get another one, he had to inform Yanhong and Liudmila that the lab soon wouldn't have money to pay them. Both waited to leave until the very last moment, as their time at Bill's lab had been pleasant and productive. Yanhong had learned molecular biology; Liudmila and Joscha had taught her biochemistry; and she finished the USMLE (United States Medical Licening Exam) in 2005 and received her green card in 2008. She accomplished everything she wanted, which made leaving it very difficult.

Both Yanhong and Liudmila went for interviews. At that time, Carlos, a new department chairman, arrived. He brought lots of money with him, and soon Yanhong, Liudmila, and even Betty from Hunter's lab found new positions. The department had three sections. One was physiology — applying flow cytometry, confocal and Calcium imaging. Another, genomics involved doing knockout mice and mutations. Liudmila went into the third section, biochemistry, where people would isolate proteins, etc. Yanhong transferred to physiology, into a nephrology laboratory in the Department of Medicine. She kept working on the fifth floor but in a different part of the hospital, in the medical department. Nephrology was the field in which she had earned her PhD. When Bill asked her about her future plans, she told him that she wanted to go back to medicine. Bill said that she would need to pass the US exams then, and Yanhong replied that she already passed USMLE steps 1, 2, and 3 CS (Clinical Skills) while she worked for him. Bill must have thought, "When did she do that?" And, indeed, Yanhong worked very efficiently at her job in Bill's lab; when she got home, her second shift at night started.

Yanhong's second shift was her preparation for the medical exams. She was well connected to former Chinese classmates who practiced medicine, and they frequently talked on Skype. Yanhong decided not to study by herself but to join a prep class at Kaplan. There, she met Shanti and Leela, two Indian ladies, and the three formed a study group. Shanti's mother was a physician, too; due to moving, she had taken the board exam seven times in seven different countries. That meant it could be done, and all three girls passed.

In the new lab, Yanhong worked for two years as an instructor. Then she became an assistant professor. Had she stayed another five years, she would have become an associate professor like

Benas, who had previously worked as a confocal manager for David. But Yanhong had other plans: she wanted to return to the practice of medicine. She was waiting only for a chance to do so.

When I returned to Chicago for visits, I not only stayed with Yanhong, she also came to pick me up from O'Hare. One time, she couldn't find parking, so she drove to a line where everybody told her that no passenger would come out. But, as if to prove the opposite, I soon exited just from there, and we met right away. "Scientists!" we heard one guy commenting about us, seeing a sticker of the U of C at her car. Another time, I was still following the baggage claim signs in O'Hare, when from out of nowhere, Yanhong appeared. Maybe scientists function alike. With Yanhong, nothing was ever complicated, and, like Charles or Shuang, with whom I shared housing for a while, she would give straightforward and very practical points of views.

Over all her years in Chicago, Yanhong had stayed in close touch with Tanner. After some time, they started to think about living together. When Yanhong visited his family, his mom could not pronounce her name, so she called her "Yanna". And she told Yanhong about Tanner. When he was two years old, Tanner hardly spoke. One day, he climbed up a chair and saw himself in a mirror. When he looked at his mirror image, very proudly, he announced: "beautiful!"

Tanner was a big fan of the Boston Red Sox. All Yanhong knew was that he loved baseball. When, for the very first time, the Chicago White Sox won, Yanhong became very excited and lined up to buy him a White Sox T-shirt. Tanner looked at her in disbelief and said: "You really don't know me!" But as good hearted as he is, he didn't want to hurt Yanhong, so he wore the shirt — but not in front of his mom.

Then came a time when Yanhong considered buying a house in Hyde Park. She found one she really liked. Tanner offered to sell his house in Toronto, and he went to talk to the agent. With Yanhong's plans to reenter medicine and possibly to move, buying a house at that point turned out to be rather a burden. Her brother said, "If you stay, buy now. If you are going away, then don't."

So Yanhong decided to postpone buying a house in Hyde Park.

Because of her strong focus on career, Yanhong rarely went on vacation. Once, she and Tanner went to New Orleans. Tanner bought a ghost in a coffin as a souvenir, saying it reminded him how fragile life is. Indeed, Tanner seemed like the perfect match for Yanhong. Having been her professor, he was more experienced, seemed wiser and could advise her well. In turn, she seemed the perfect match for him. As he said, he could only love a smart woman.

Yanhong had been applying for a green card, but she was too proud to get it by marriage. She entered the green card program as a Chinese national; later, she learned that she should have applied under her Canadian citizenship, as it would have been faster. Once the application was placed, it could not be changed, so Yanhong had to wait a long time.

When Yanhong was reflecting on her life, she said that she was very lucky. Lucky to grow up in a family with love. And lucky always to find someone to work with. She said she considered it a privilege to work in a good job that she enjoyed. Every year, she sets a new goal. That way, life never becomes boring. It took time to adjust from China's countryside to the rich city of Chicago, but Yanhong mastered the challenge quite well.

Out of the three-hundred-and-sixty students who graduated with her back in China, eighty now live in the US — about 22 percent. Yanhong still stays in touch with some of them. They meet every five years, and Yanhong once joined them in Washington, DC.

Of course, bad days come along, when experiments don't work for no plausible reason. But Yanhong considered life as gift from God (leaving the definition of God open). She said, "If something bothers you, make it not bother you anymore." During my last visit to her apartment in Chicago, I remember being puzzled about a friend's behavior. She simply said, "Why bother? If she steals your time, just forget about her. Life is too short to be wasted!" I followed her advice and disconnect to people now, if needed.

LIUDMILA

Some years after Yanhong joined Bill's lab, Liudmila came to the lab, replacing Joscha as a research associate. She, like Katarina, came from Russia, but she came from the countryside. Tall and slim, Liudmila had beautifully dyed reddish hair. Bill simply gave Liudmila's CV to Yanhong and asked her what she thought. Liudmila had worked on rice, and Bill said it was all the same to what they were using. He said that Liudmila could usefully apply the methods she knew. So Liudmila started to work on Trp, microarray.

Liudmila had a husband and a teenage son, Juri, back in Moscow. And she would do anything for them. Her life in Moscow had been very hard. Once she told her story about being pregnant with Juri and having to go to the hospital. Some people in power there were not nice under the ruling communist party. Liudmila was used to the harshness of the communist regime, which was all about giving orders. Highly pregnant with Juri, Liudmila was hardly able to walk at the hospital and was required to cross the room with a thermometer under her arm, which kept falling down. The nurses would yell at her each time it happened, expecting her to pick up the thermometer, she was not able to, which caused even more yelling. When the baby was born, they had no clothes for him. People got purchasing permits, which allowed them to buy, but they simply got what was available. In order to get clothes in right sizes, people would have to exchange things until they got what they needed.

Liudmila had come to the US so Juri would have access to a good education if he came after high school. She started in the Anatomy Department, but got very sick and had to undergo spine surgery. Still, Liudmila bravely managed.

In Bill's lab, Liudmila worked closely with Yanhong. Bill needed two people in his lab, one for physiology, Yanhong's role, and one for biochemistry, which Joscha had done before. Now

Liudmila would do all the biochemistry, and she was excellent at it. She often suffered bad back pain from her spinal surgery, so Yanhong would help her whenever there were heavier things to carry. Liudmila was also well connected with other Russians on campus. One time, she asked one of her friends to lend me a bike so I could go biking with Yanhong. That Saturday, I went with that bike towards Yanhong on a straight street; I was going through an intersection on a green light when a car turned from the left and hit me. The bike and I were thrown up in the air, while the car passed. Yanhong, who was waiting for me just a few meters away, witnessed everything. When I landed on the street, I didn't move. I was too afraid of pain. In addition I also feared getting run over by oncoming traffic if I would sit up. Since I didn't move, people nearby expected me to be dead. I heard them screaming, and people reached forward. The car driver, a teenaged boy from the poorer neighborhood, and his pals even failed to stop after the accident. Yanhong took all precautions and informed Hunter immediately. She also told Charles, whom I planned to see that night. But, miraculously, not only was I fine, the bike belonging to Liudmila's friend was fine, too!

The Fifth Floor

Liudmila had a golden heart for people and animals. When Yanhong moved from Chicago and everything was packed, she realized she had no place to sleep. So she called Liudmila at around 9 pm, and Liudmila kindly invited her over and even cooked for her.

Liudmila had a beautiful angora cat, Persik, the "king." In her bedroom on 54th Street and Ingleside, she also had a giant fish tank with just one fish, Gorsha, the "professor." Gorsha was very special. Liudmila had adopted the big fish tank with several hand-sized blue fluorescent fish in it, but Gorsha killed them all. Now he was the only one left. Once, when I stayed in Liudmila's place and Liudmila had left the bedroom to me, I would wake up in the morning with a fish staring at me; knowing Gorsha's background, one can get mixed feelings. Liudmila also had several smaller fish tanks in the living room; she was quite knowledgeable about animals.

One time, the warm-water fish in the smaller tanks got Ichthyo, the "dot disease." Liudmila had tried treating them with chemicals, but the disease didn't go away. So we caught all of the fish, and I treated fish after fish by hand in a bucket. It worked well, but we two scientists had not paid attention to the water temperature in the bucket, and the first fish left rather drowsy.

Liudmila also took in a friends' dog for a few days, and she adopted Katarina's cat, when Katarina had to give it away.

As her family was in Moscow, Liudmila sent them all her money. That was the plan, she came to the US to make money. They had planned that her husband would come when Juri entered university. But, not only that Juri would prefer to stay in Russia and study there, through the Russians connection via Katarina, Liudmila found out some unacceptable things about her husband, so the marriage didn't last. Poor Liudmila was now alone.

That spurred Liudmila to give her life a new direction, and she started looking for a property to buy. The U of C was working to make living south of 61st Street more attractive, and Liudmila bought a beautiful apartment at 61st Street and Ingleside. Because Liudmila was a very talented tailor, she made all her own curtains, cushions, tablecloths, etc. And she offered me clothes, too. She gave me the precious dress she once made for her thesis defense in Moscow. When I was about to leave on the road trip, Liudmila presented me a very nice pillow she made with a patchwork pattern.

Liudmila was also a very talented gardener. In addition to her apartment, she rented a garden and started to grow her own vegetables and flowers.

The real estate agent, Dima, who had found Liudmila the apartment, came from the Ukraine and therefore spoke Russian, too. They soon became friends and more and ended up getting married. Dima was a very nice, open guy who could just talk to anybody. He was short but seemed very lively and active. He had a daughter, a lawyer, whose husband was a lawyer, too. Liudmila and Dima would visit their children together from time to time. And, since Dima had a green card, Liudmila's stay from now on was secured. Liudmila would sit and smoke outside on the veranda, while Dima preferred to exercise. Everything seemed perfect.

The lab's routine was for Bill to arrive at 8 am, for Yanhong to arrive an hour later, after her morning swim, and for Liudmila to come late and stay late. This open schedule for everybody to chose when to come and how long to stay made people really enjoying their work. Back in Basel professors like Zeus would insist on arriving at 9 am or others even at 8 am, which does not work for everyone. Liudmila was very good technically and easy to get along with. Since Bill was an early bird, coming in early and leaving at 5 pm, he had to learn some flexibility. Yanhong learned a lot about real estate from Dima and considered planning to buy a house, too, long time before planning to live together with Tanner.

Shortly before my final return to Europe, I had to store a few things. First, I left them in Hunter's basement, but because it was easier to get to Liudmila's place, I soon decided to move most of them from Hunter's basement to Liudmila's basement. Yanhong did all the driving, and Liudmila and I started shipping several boxes back to Europe. We went to my favorite post office on 61st Street, a few blocks away, and we were moved by the kindness of everybody there.

While Liudmila unloaded the car and I tried to get the parcels into the post office, everybody gave us a hand. It was considered a "not-so-safe neighborhood," but I always preferred this place over the post office right across from Abbott Hall.

Then Dima got sick. He lost his job and had to undergo kidney surgery. Fortunately, Liudmila was there for him, and after recovery, he resumed working. The next hardship was when Bill as mentioned earlier lost his grant, forcing Liudmila to find another job. Liudmila began working in another lab but when it came to publish the results it seemed that her results had been changed. She went to complain. What followed was the worst nightmare a scientist can face, to argue with the boss about results. As she did not want to stay with her boss and not to see him around anymore, she left not only the lab but the U of C. Luckily, she found a nicer job in no time that even paid better. In her new position, she became "the Western blot specialist." When I visited Liudmila in her new lab downtown, she told me that they no longer did gels themselves like the rest of us, including Joscha and even Dominik, used to do. She told me all the experts' secrets about Western blotting.

Yet another hardship was awaiting Liudmila: she started to develop epileptic seizures. When she went to see a doctor, they found a brain tumor in the frontal cortex. It was a glioma, which had to be removed. This time, poor Liudmila had to undergo brain surgery. Dima was a great help. Everything went well, and Liudmila returned to work. Liudmila told Yanhong that things like that really change your life: you never become angry over little things again. Liudmila and Dima still enjoy their life, living with their cats, seeing their children, and going on vacation from time to time.

Over the years, it became more difficult to keep seeing everybody because they had moved to work in different places. I would take the shuttle bus from the Fermi labs to see Joscha at the Argonne labs or Liudmila downtown Chicago. At least, that was still Chicago area. Later people would move further away within the US or even overseas like myself. Timing prevented us from all getting together, but we always tried to meet up with everyone from our group in Hyde Park.

NAOMI

Naomi worked for Bill as a technician while she studied film. She was from Belarus, rather quiet, shy, very gentle and Jewish. She wore dark rectangular glasses and beach shoes. She had wealthy parents, though she never showed it. Yanhong once said, "She is very bright. She works fast, and does whatever you ask her."

Bill always liked to have two students working for him. Naomi worked with Seo-yun, the professional dancer. Later, Pjotr and Chang would take their places.

Naomi had a little white dog who came to her as a rescue. Before coming to Naomi, he was treated very badly, so he would not allow other people to get close. Once, when we were driving together, Naomi asked me to ignore her dog on the backseat as he would be fearful of others.

When I followed her advice, her dog amazingly looked for contact with me, and we became friends. Up to then, only Naomi and her boyfriend Bassem had been allowed to touch him, and I felt very honored.

Bassem, from a wealthy family from Lebanon, applied to medical school. He asked me for a letter of recommendation. In one of my moves, I left a few furniture items with him, including a favorite chair and a lamp. Naomi and Bassem did not stay together and sadly, I never saw my belongings again.

Once, when I was looking for a place to live, Naomi offered for me to rent one of her roommate's rooms. Naomi and three other people lived in their apartment. Two of them, a couple, simply moved into one room together while I lived there. It was a new experience to meet her roommates, all very young people of different backgrounds. One girl studied art and dreamt of going to Italy. She invited me to one of her violin recitals for which she had practiced hard. Her boyfriend, whose room I now occupied, was a math student, who seemed to be not very successful. The third had been studying medicine. However, since Naomi was not very fond of him, I didn't talk much to him. Living in that apartment was like studying social sciences, observing their interrelationships and how they managed to take care of their studies and household requirements. Naomi's dog liked to visit me in my room, and Naomi would show me pictures she had painted for movie sets. We would occasionally drive to a pizza place located up north; some years later, Vilma moved to a place just around the corner from it. And Naomi would always bring one pizza home for her little dog.

When I asked for contributions to this book, Naomi kindly wrote, "I remember working at the U of C hospital very well. It was my first job on campus. I had moved there from San Francisco less than one week before my interview with Bill. It was a great group of people to work with. Hunter and his lovely team were just down the hall, and we would run in to each other frequently. I would work in between classes, keeping things tidy, and helping to prep for Yanhong and Joscha. Then Lilly joined our lab, and Liudmila. I miss all those great folks! I remember Lilly. I also went to the movies at Doc. What a great little theater!"

Indeed, this theater was something of a kind. Next to the International House and Ida Noyes Hall, the theater showed movies which were hand selected and only played once, as I remember. It was always a great atmosphere in the one big room they had. For the latest Hollywood movies one had to go downtown to the big theaters.

Later, Naomi moved to LA. When Debra and I went on our road trip, we not only visited Naomi in LA, but we stayed with her at her dad's house in San Francisco. Her parents were divorced and her dad had remarried and lived with his new wife in the neighborhood of the real Robin Williams, right across the Golden Gate Bridge from the city. Her dad was very kind, and it was a great shock when, years later, he was diagnosed with cancer in a final stage. As I worked in neurosurgery his wife sent me the scans his doctor had ordered, and they decided not to go for treatment at the Anderson Center in Houston, where Vilma worked, but to stay at home. Sadly, he passed away quickly.

EDWARD

Writing about Edward and James has been the hardest, as both of them had left us way too early.

Edward, simply called "Ed" by everyone, was one of the most remarkable people I have ever met. Bearing a resemblance to Sigmund Freud, Edward was born in Chicago on the Swiss national day, which already made him special to me. At age twenty, even before his graduation, Edward was diagnosed with *dystonia musculorum deformans*. That meant a similar life to that of Steven Hawking, the well-known physicist, fighting uncontrollable muscle contractions, speech impairments, and confinement to a wheelchair. After earning a BA in liberal arts and a BS in biology, Edward decided not to go on to medical school, as he had planned. He went into neurobiology, instead. He got his PhD in biopsychology at the U of C and married Michelle, whom he had met six years before and who became an MD herself.

After graduating, Edward went to Gothenburg in Sweden for a postdoc. There, he worked with Dr. Carlsson, who shared the Nobel Prize for medicine in 2000 with Hunter's former PI Dr. Greengard and Dr. Kandel, whose lab I had the honor to visit in NY later. Edward did another postdoc at Stanford before returning to Chicago, where he stayed. In 1977, he became a tenured professor and not only kept on teaching but also continued to sail, despite his impairments. Among his many awards, he received Chicago's Golden Key Award and an honorable award in Gothenburg from the king of Sweden; we always wanted to watch his video of this event, but we kept putting it off until the next time.

I met Edward because Hunter, his friend, was looking for a room for me to rent and asked him. Edward had already two other students living in his big house and I moved in, too. This practice that professors would sublet rooms to students was another amazing setting, professors in Europe hardly ever do.

Edward and his wife Michelle lived on Woodlawn, along with several nurses who took care of him on alternating shifts. Edward and Michelle's three children were grown up and no longer lived with their parents. They already had a grandson named Eddie. Edward would often talk with me about them — about his daughter, a veterinarian, and especially about Ian, who, like me, had a pilot's license.

Their house was really quite big. Due to previous burglaries, they had installed a sophisticated alarm system that was directly connected with the police. In order to move from room to room, a security code had to be entered at each door- and the codes were changing every week! Because I found this rather inconvenient, I did not stay very long. Now, since the alarm system was crucial, Hunter did not have the heart to tell Edward the truth about my decision to leave. He said that I was allergic to his cats. But Edward was such a warm-hearted person that he seriously considered giving the cats away!

The Fifth Floor

When I moved out, I promised Edward to visit him from time to time at home or in his office on the first floor and I did.

Edward's secretary, Joyce, still worked for him and was just as welcoming as him. Whenever I went to see him, he gave me the feeling there was nothing greater than to see me. He was one of those very rare people who never found it inconvenient to meet anyone. If he happened to be bedridden when I visited him at home, he would still smile and invite me in. Not being a native speaker of English, I was particular grateful that I could mostly understand or at least guess what Edward was trying to say as the disease had a great impact of his speech. And he always had so much to say!

At home in his office located to the left of his front door, he would be excited as he explained how the speech recognition software on his new computer system expanded the range of things he could do. Or we would sit in the living room, to the right side of the front door, where he kept a piano; usually I would not perform in front of others, but I would play a little on rare occasions just for the two of us.

We were both interested in toxicology, and enjoyed travel. He liked to fly Swiss Air, and we would fantasize about going to Israel together. So it was from Edward and Hunter, I had learned all about the Weizmann Institute in Israel and I dedicated my PhD thesis to them.

Though Edward was severely handicapped, he never gave up. He would take the elevator to the fifth floor in his wheelchair to see me. And like Hunter Edward was despite of his confinement all passionate about science. It was just such a great atmosphere I never experienced anywhere outside the US or Israel. In order to be more independent, he got a special driver's license that allowed him to drive with a set of seven mirrors. We all admired Edward, and I often talked with Yanhong about him. But sadly, and shocking to all of us, his independence almost proved fatal: one time, when he drove to Abbott Hall, somebody knocked him down and stole his car, which was found later totally burned out in one of the less safe neighborhoods north of Hyde

Park. The police reported that he had become a victim of rival gangs, and it took Edward quite a long time to recover from the shock.

When I was away from Chicago, Edward sent me the nicest emails ever. Later, I learned that these would take him hours just to type. He had to type them using a stick attached to his forehead, as he had no control over his hands because of the spasms caused by his illness. As was his way, he began his emails with "Dearest Lilly" and ended with "fondest regards," "love," etc. Because I thought this must be the normal American way, I applied it to my own correspondence with everybody on the fifth floor. It didn't take long for my friends to start teasing me about it. And instead I learned to close with "kindly," a term we jokingly used for best friends.

Edward was the best friend one could wish to have on one's side, in particular if someone turned sick. He spoke openly and honestly about his own limitations and frustrations, and nobody showed more joy than he did when everything went well. Edward was one of those very rare people who would always meet you with complete joy and open arms and never ever disappoint you.

During that time, I introduced Yanhong and the two ladies, Liesel and her cousin Ursula, to Edward. They were deeply impressed. All of them were stunned by his immense kindness and joy. Yanhong said that, behind his disease, one could see his bright mind. That was correct, and he would always talk to me about my projects.

I would visit him in a hospital up north, once he had to undergo surgery. In that hospital, they attached the patient's name onto the front of their door, and by mistake Edward's name plate said MD. Much to his amusement, many people came to see him, assuming he was the doctor on duty.

When Edward died at the age of seventy-two, in 2007, Hunter emailed me. We all deeply felt the loss of this most remarkable, warmest friend.

JAMES

Just before I moved out of Edward's house, Hunter called James to see if he had a place for me to stay, and James asked us to come over. So, late one afternoon, Hunter and I went over to James's place, just a few houses down from Edward's on 55th Street at Woodlawn, more or less across from where Yanhong lived. It took James awhile to open the door, and he just said: "I'm eating dinner right now!" We apologized, and he let us in. To my great surprise, James' furniture and interior decor looked remarkably like I had in my apartment back in Basel. It turned out that we had purchased everything from the same stores in Paris, and if it wasn't exactly the same, it was very similar.

Like Edward, James was one of the nicest persons I ever met. And like Edward, James was

Jewish and very kind and gentle. James even was at that time the youngest social science professor in the US. When I once asked him about his age, he, like Hunter, didn't tell me the truth. However, where Hunter had made himself younger, James made himself older.

So Hunter and I looked at the room he had to offer. It was very small and almost empty, and James set the rent rather high. But Hunter said he could bring me some bedding, so I agreed to take the place. While Hunter left to pick up the bedding, James asked me if I'd like some of the broccoli he was eating, and we had our first dinner together.

Very soon, we became the perfect roommates, and we had a routine. I always got up first in the morning and showered. The bathroom, more or less across from my room, had a Mickey Mouse motif around the shower and a hippopotamus soap dish, as James loved hippos. By the time I had dressed, retrieved the newspaper from the front door, and prepared our breakfast, James would get ready. Then, we ate and read the paper together. On Saturdays, James usually baked a cake. The first time, I started laughing, because he tried to make a chocolate cake by melting the chocolate over the fire. I showed him how to melt chocolate in a water bath; from then on, we always baked the Saturday cakes together. Just like on weekdays, I would return home at about 6 pm and prepare dinner. We would eat together, talking about the day, and then I usually returned to the lab until around midnight. James would stay up, waiting for me to return. Then, he would go to sleep right away.

Since we had no clothes washer, we had arranged to do our laundry upstairs in the landlord's apartment. It always felt awkward for both of us to interrupt these people's life by doing our laundry, but there was no other way.

James loved his apartment and had many memories associated with it. After receiving his PhD from Harvard in 1990, he moved to Chicago, and this was his apartment. When he left for a limited time — to the West Coast, I think — he sublet the apartment in order to keep it, though the rent was rather high. While the building didn't look very special from outside, the apartment was huge and very nice. I could easily understand why James had kept the place.

On Saturdays, he would also call his former girlfriend, with whom he used to live there. They shared the apartment for a long time as a couple, until she moved away for another job. We didn't talk about it, but James obviously missed her very much. He still kept pictures of them everywhere. Later, when Kelly (his new girlfriend who later became his wife), came to visit, I would discretely put the pictures away.

Sometimes Hunter would call for me, and James would simply say I wasn't there. I didn't know about this until the next time I saw Hunter, who would ask me where on earth I had been. When I asked him why, James told me that I was working too much and that he had simply told Hunter I was not at home.

Then James asked me if I would like to play squash with him. I had no idea about squash, so I secretly took a few lessons to avoid looking too stupid. Later, I found out that James just had just done the same thing. We could have simply started together.

James also asked me if I would like to go sailing with him, but somehow we never did this. We occasionally ate lunch together at the Social Science Reseach Building, where they served excellent hummus, and we would talk about our day, the news, and the French people and history, he had been writing about. I bought a little book about social sciences at Powell's bookstore to become a better conversation partner. On his birthday, on Columbus Day, I would place a bunch of flowers on the breakfast table, and we would spend extra time talking, since it was a holiday anyway.

One morning I went into the kitchen, and it looked to me like we had a burglary. After I called James, we found that a squirrel had gotten in, biting through the metal wire at the kitchen door and scattering food from boxes on the floor. We both laughed as we cleaned up the mess.

Though James was very successful, he never seemed conceited about his many achievements. The first time I met Kelly, his later wife, I asked her what she was doing or planning to do. She openly replied, "Nothing, and I don't know!"

James, on the other hand, always seemed goal-oriented. Once, he came to visit me on the fifth floor, surprising Hunter as well as others, like Raj, who wondered why the well-known social science professor was paying us a visit. I gave James a tour, had him look through the microscope, and introduced him to my friends. James asked very smart questions, which he continued even after we got home.

Once in a while, he would ask me to watch a movie with him, and he would bring us a plate of grapes to the couch. He would give me his jacket if he thought I was cold, and he offered me his car whenever I wanted. James was a wonderful friend.

Each time, even for short trips, I had to leave Chicago, James gave me a long hug. I always stayed with him as long as he lived in this apartment. But things were about to change for him. Sought out and followed by admiring, annoying teenage girls from his classes, James no longer wanted to live in Hyde Park. By then, I was doing all of our grocery shopping alone. He moved to an apartment in Wicker Park, in Chicago's north end. He asked me to move there with him, but Wicker Park was quite far from Hyde Park, which would make it difficult to get back to the lab every evening. Although he offered me the use of his car, I decided not to follow. That was when I got my own apartment on 55th Street and Blackstone.

We still occasionally met for lunch in Hyde Park, and once I went to his department to pick up a box he had respectfully kept for me. The next time I saw him, James didn't look good. I remember mentioning it to Hunter, who was also worried about him. James and I would talk openly, but there was no immediate indication that he needed his blood checked, which in hindsight might have been the right thing to do. Sometimes when we sense things, we should pursue further.

Later, Kelly came to Chicago to live with him in Wicker Park, but she didn't like Chicago. In 2000, they moved to Yale, where James started a new life. He and Kelly got married, and Kelly earned a degree in psychology. Sadly, James had developed a very rare kind of leukemia. He bravely underwent treatment, and no one expected anything bad. So he went to New York, where his mother lived and where he had a position as a Visiting Scholar at Columbia University. Then rather suddenly and as a shock to all of us he passed away in 2002, at the age of only thirty-nine. We had just lost our first dear friend, followed by Edward in 2007.

VILMA

Vilma was the sort of person that immediately makes you feel comfortable and taken care of. I met her on my very first day of working for Hunter. A student in Hunter's lab and of Lithuanian ancestry, Vilma was a tall and international girl: she was born in Brazil, schooled in Germany, and grew up in the US. She spoke at least three different languages fluently and could easily switch from one to another.

I was still under shock from my experience with the graduate student in Lynn's lab. Having been an excellent student always before, this experience sent me at the edge of having lost my faith in science. Vilma took care of me and became immediately my mental bodyguard.

We met under rather exceptional circumstances. Hunter had brought me up to his lab on the fifth floor to show me around. When Vilma came in, because she did not expect anyone to be standing in the door, she spilled the ethidium bromide gel she was carrying - and it landed right on my shoes! Ethidium bromide is pretty toxic. She tried to pick it up, despite its toxicity. In order to rescue the experiment, Vilma simply called out, "DON' T MOVE!" And we both laughed.

Vilma and I soon became best friends. Like Yanhong, Vilma could teach me a lot. She introduced me to new things in the lab and showed me around the campus and downtown. Vilma's parents lived in the western part of Chicago, and when Thanksgiving approached, she invited me home. Of course, I had no idea about Thanksgiving at first. This was another American custom we don't have in Europe. Vilma's sweet mom could speak German too, while her Dad and younger brother did not. It was clearly the ladies ruling with languages. We had a wonderful, big Thanksgiving with many typically Lithuanian dishes. On TV was some sort of reality show with family clashes, I had never ever seen before. All was as exotic as it could get.

The grandmother of Vilma on her mother's side lived in Heidelberg, Germany, and one summer Vilma went to visit her. We would meet while I was in Switzerland and together we went to visit Zeus in his lab in Basel, participated in a special seminar, took pictures and even had cake with him.

Vilma was very sociable, never to upset, and practical; and she was involved in all kinds of social projects. At that time, she volunteered at the children's department at the hospital. Soon she got me volunteering, too. I still remember when, one night after work, I went to see a little four-year-old African-American girl, a patient who was awaiting a liver transplant. When the little girl saw me, she called out: "Oh, MAM, you look sorta tired. You should gone home first and got some rest, man!" I looked at her, to make sure she was really only four years old and not seventy-four.

While I went to visit more little patients from time to time, I met a Japanese postdoc, colleague of Sanja, from Warren's lab on campus. His wife Miyu, also Japanese, had been a children's nurse.

Because of language problems, she was afraid she would never again get to work with children. Since I speak some Japanese, I arranged for her to come with me soon. Miyu and I became good friends, and we spent some time together, outside of our visits to the Children's Department. One time, Miyu invited me home for lunch. I was amazed to find that they had two refrigerators. When I asked her about it, she said that one was for the food and the other was for her husband, who worked on cholera. You never know what scientists keep in their homes!

I also helped sometimes in the ER, if they were short-staffed. As I was not an English native speaker, I did not always get what patients said to me and I had to rely on what I saw. Shootings occurred every weekend in the poor neighborhoods near Hyde Park, and when I asked the injured what happened, it was only for the purpose of letting them talk, not that I had a chance to understand everything. In turn, I would give them orders. "All right," I said, "tomorrow morning you come back. And you really have to be sober, understood?" "Yes, Ma'am!" came the answer, with a respectful look. What I actually meant was that they had to fast — not eat anything. I was not talking about alcohol abuse, but my dictionary said "sober" instead of "fasting" and they looked at me as if I knew more than they wished.

Vilma also took part in a project that renovated houses called "Habitat for Humanity" and in a project at the state prison. Over the weekend, she also taught at a Lithuanian school in her parent's neighborhood. When I was an undergraduate student, I hardly had time for other activities, but Vilma seemed to be involved in just everything and did an excellent job. Vilma was just amazing! She lived most of the time in a basement apartment between 53rd and 54th Street on Woodlawn that she shared with Jenny, a student at the law school. The two girls were very open and great friends. Jenny's fiancé used to work in the hospital and had a grant for an idea he had come up with as an undergraduate. Something like that would never happen in Europe where hierarchy rules, and this was quite impressive. It made work much more appreciated and was an incentive to make own suggestions and approaches.

The Fifth Floor

When my birthday approached, Vilma asked me if I had a wish. Not knowing that "quark" (a sour skim milk product) is unavailable in the US, I had asked for the impossible, a European style cheesecake. Poor Vilma did everything she could to bake the cake, which still embarrasses me today.

Vilma taught me many things about the US that I had no idea about. One time, while talking about yogurt, I asked her why in the US they mention every single thing about the product but the bacteria strain. Vilma said that if they used the word "bacteria," nobody would ever buy yogurt again. Later I could see what she meant. I was looking for melatonin in a Walgreens store and asked about it. The salesperson said, oh, that is a vitamin — and sent me to the vitamin section. Of course, melatonin isn't a vitamin, but a hormone. If they'd call it a hormone, people might not want to buy it.

In 1999, NATO bombed the Chinese embassy in Belgrade, and a huge debate occurred in the media about whether it was a mistake. At that time, responding to a call for short stories for a Chicago Literature Award, I started to write a story related to this event. As I didn't expect more than to participate, I didn't show the story to anyone to review and simply sent it in. A few weeks later, Hunter got a call in his office. When he heard something about literature, he simply hung up, shouting, "Wrong number!" Well, they called again, this time asking if Lilly worked there. Yes, she did, but …Before he could hang up again, they told him that I had won the first prize. Half disbelieving, he sought me out. Feeling like I had been caught *in flagranti*, I admitted to participating in the contest. Hunter frowned and suggested that I should really focus on science.

Now, the prize would be awarded at some kind of ceremony, but as a non-US citizen I needed help. Again, Vilma stepped in and handled everything. My short story got printed in the campus newspaper Maroon, which we would pick up at places like Reynolds Club, and the page with my poem was hung up in Hunter's lab.

Years later, when Scarlett came to work for Bill, I introduced her to Vilma. For a while, Scarlett would hang out with Vilma's brother, but that didn't work out. Vilma used to tease her brother that things worked out better for Scarlett apart from him.

After working for Hunter, Vilma moved on to work for a lady professor, whose name I don't recall. She not only worked in the lab, but also babysat her new PI's young son, who was autistic. Since I'm very interested in autism, Vilma asked her boss to invite me to her next party. I went with Vilma and indeed found the boy surprising. Not only could he as a toddler change his own diapers and carry two one-and-a-half liter bottles at once, but he would quote long passages from children's movies and advertisements in place of personal communication. The insight I gained from this family taught me far more than many textbooks ever could.

This little family was quite amazing. The husband of Vilma's PI was working with a gas corporate and had moved from East coast to West coast and back. In the US people would easily buy and sell their homes and not be as restricted to one place as people tend to be in Europe.

After Vilma earned her Master's degree, she decided to become a physician's assistant. She moved to Houston, where she worked with cancer patients — a position which couldn't be more perfect in benefiting patients.

When I tried to collect memories for this book, Vilma kindly replied. "I remember him [Hunter] actively playing soccer [...] and I remember being initially intimidated by him and just hoping that I did a good enough job washing the lab's glassware. I guess he saw some promise in my dish-washing abilities, because he soon took me under his wing and invited me to work on the dynamin (a GTPase responsible for endocytosis) project. He awakened in me a new love for research. He opened the door for me to participate in the lab; he taught me to be meticulous even in what may appear to be a "menial" task at first glance, as those details are important in running a successful experiment. He was the first to expose me to scientific literature and helped train me to read scientific journals with a critical eye. Although I no longer work in a lab, I have assimilated those skills into my work taking care of patients with advanced kidney cancer."

BETTY

Betty came to Hunter's lab as graduate student. Though she worked in Hunter's lab for years, she did not go on with graduation. After working for Hunter she worked also for Amelia on the same floor and became at some point what everybody called "the confocal queen." She was one of the few "American-born" members of the lab, and she could understand and comment on political or sports issues that most of us foreigners didn't understand. Her step mum would occasionally come to visit her over the weekend and she was good friends with Pam and later Katarina. Betty was a big cat lover and kept two cats. She was rather short and loved ballroom dancing, she went to practice for all dressed up on the weekends. As mentioned before, Betty had a very social side and helped me when I fell sick by the scorpion in Mexico.

PAM

Pam, whose name was actually Betty Pam, was another graduate student in Hunter's lab. She was very quiet and like Betty she was American-born too. Conveniently, Pam loved cats, too and they took care of each other's cats if one was out of town. Pam also had twin sisters as siblings. They were much younger than her and occasionally came to visit. Pam successfully defended her thesis and went on to Harvard Medical School, where Debra and I met her on our road trip.

RAJ

Raj got his Master's degree and worked as technician for Hunter. Of Indian descent but born in the US, Raj spoke perfect English. He lived with his parents and younger brother off campus. His mother was an Indian from an Indian community in Mombasa; later, when I went to Kenya once, Raj asked me to take pictures for him. Raj was a very nice guy. He could be skeptical of other people at times but also very outgoing at other times. He made no secret of the fact that he was gay, and he kindly invited us along a few times to join him and others downtown. So, for the first time ever, I went to a gay bar. I don't know what I had expected, but it was a nice, rather expensive place with a grand piano. To Raj's disappointment, it was too early, so almost no other people were there. However, for me, it was still nice to see.

Once, I remember Hunter talking to me in the room with the huge centrifuge in and the door closed. The door had a small vertical window. Raj was walking along the fifth floor calling out loud for Hunter: "Hunter? HUUUUNTER!" When Raj approached our room, Hunter suddenly got on his knees. A little later, Raj asked me through the window, "Lilly, have you seen Hunter?" Now — with Hunter on the floor gesturing at me not to tell about him and, at the same time, being a friend of Raj — I found myself in a real quandary. What should I do? I suggested that Raj look for Hunter later and started laughing about Hunter, who was still on the floor.

As customary for Indians, Raj as the older brother was supposed to get married first. When his brother started to grow up, he urged Raj to tell his parents he was gay, in order to be allowed to marry. So Raj told them. Sadly, Raj's parents didn't take it very well. Not long after, his mother passed away from breast cancer.

One day, Raj had to fly some place in the US, and he went to get flight insurance. Curious, I accompanied him to an office above the coffee shop across from Abbott Hall. He bought insurance naming his brother as the beneficiary in case of a plane crash. I thought he was really very kind and caring about his brother.

At Christmas, Raj bought Hunter a nice gift — a clock, he told me. He wrapped it and placed it on Hunter's desk. When Hunter saw the box with his name tag on it, he nervously called me in to ask about it. Because it was ticking, Hunter feared the worst — he thought that it might be a bomb!

Raj went on to UCLA, where he earned his PhD. Debra and I met him in LA on our road trip, and Raj showed us the lab where he was working with *Drosophila*. Later, he moved to Seattle as a postdoc. When his boss in Seattle decided to move to Heidelberg, Germany, Raj joined him. We met a few times in Heidelberg, where I later worked at the same institute, the DKFZ, as he did. We would have lunch together as in old times and I came to see him a few times in his PI's lab, where they kept a beautiful by now multichannel confocal microscope. On campus parakeets were nesting here, too, but it just wasn't the same as Chicago, missing the social life we were used to.

KATARINA

Later in Hunter's lab, Katarina joined us as a postdoc. She came from Moscow in Russia, where her father had been a medical professor. Quite tall and blond, she dieted for the entire time I knew her — always eating very small lunches to keep her figure in shape. Indeed, she could have been a model. Her cosmetics were always perfect, but she was also a good scientist and an excellent support for Hunter. Katarina was a close friend of Liudmila, who also came from Russia.

Katarina was married to a well-known Russian artist. They had met back in Moscow, when Katarina asked him to paint her portrait. Sadly, his kind of art was not much appreciated in Chicago. When he couldn't continue his artistic career, they started to have problems. In

Illinois, one can file a request for divorce without their spouse's knowledge. Katarina decided to take this step. When her husband found out, he threatened to show up at the lab and cause a massacre. Hunter had to inform campus security, and we ended up working under raised alert for a time. Fortunately, time passed, the divorce became effective, and no revenge was taken on the fifth floor.

After her divorce, Katarina's son fell into a depression, and Katarina had to focus on him. Her son recovered and married a girl from California. For Katarina and Liudmila, who had both lived through so much turmoil in their private lives, it was time to focus on themselves. She and Liudmila became beauty experts, ordering clothes and cosmetics online. Each of them attracted a new husband- that was Dima for Liudmila- and, later, bought a house. So, Katarina remarried. She and her new husband lived off campus up north. Her husband worked for a company that had him traveling to China frequently. He asked Yanhong where to find the best presents in China; over time, this led to Katarina gaining the admiration of even the Chinese for her exquisite Chinese outfits. Sadly, her husband suffered from diabetes and hypertension, and he had a heart attack.

Katarina had a Persian cat, but her new husband was allergic to it, so she decided to give the cat away, which Liudmila had adopted.

JINGLE

Jingle came to Hunter's lab while working on her Master's degree. She was a nice girl from the Philippines. I only met her during one of my shorter stays. As Yanhong once put it, she was a "very typical Filipino. If I think of the Philippines, I think of Jingle."

Jingle was known to be very protective of the lab she worked for. In the process of getting his divorce, Joscha sometimes had to interrupt his work day for long phone calls. Yanhong, trying to respect Joscha's privacy, asked Jingle if she could use the phone in Hunter's lab to inquire something at a company. But, since Yanhong didn't belong to Hunter's lab, the doors were never opened to Yanhong.

When Jingle finished her Master's degree, she initially wanted to go on to write her thesis in Amelia's lab, but then decided to become a graduate student in an endocrinology lab. Her PI, Kang, was the first to develop vitamin D-knockout mice. Jingle handled all lab ordering; she knew all the prices and companies and her help was much appreciated. She married a Filipino and became pregnant. Yanhong met her once at the swimming pool, where Jingle exercised hard despite her pregnancy. Yanhong told her she had better be careful.

THEODORE, EDGAR, OLIVIA

Theodore, a well known professor at the hospital retired in 1998 and Hunter would take over his lab members. Theodore was born in Germany and being Jewish he had fled Nazi Germany. He later moved with his wife to Israel, where his son lives with his family. Theodore would occasionally visit Hunter on the fifth floor. Despite of his many achievements he was a rather quiet, modest person, who enjoyed classical music and even wrote poetry.

Upon his retirement Edgar, another American-born scientist and cardiologist, joined Hunter's lab. Edgar was a rather slim blond about Hunter's size. He was kind, with an even temper. His calm demeanor made one feel that the world was in order. He lived with his wife and four children somewhere north off campus.

Once, when Edgar returned from a trip to California, he told us about it. "In California," Edgar said, "everything is so nice. Even the police are like gods!" — a saying that Yanhong always quoted when talking about California. When Hunter's lab closed, Edgar left the U of C, but he still lived in Chicago not far from Hunter.

Olivia was African-American with a nice sense of humor. She lived south of Hyde Park and, by some secret, never showed any signs of aging.

She was a technician and actively attended church in her neighborhood, and one time she invited us to a play she was in. Of course I went, and Olivia kindly drove me. As the only non-African-American there, I was a little afraid of sticking out, like I had been before in Elijah's church. Olivia had to leave me on my own while she went to get ready for the play. A little girl, probably not much older than eight, asked if I wanted to sit with her. I felt like I was eight years old, too, and we hung out together for the entire evening. My new young friend told me that her grandmother raised her. When she learned that, like Olivia, I worked at the U of C, she said that she wanted to study hard in order to become a scientist, too. After that, I always asked Olivia about her.

Another time, I accompanied Olivia to a baseball game — the White Sox against the Texas Rangers. Olivia, a big White Sox fan, not only knew all the players but knew many of the people around us in the stadium. For my part, I didn't have even the faintest idea about professional baseball. Whenever I thought it was time to applaud, no one else did; but when nothing seemed to have happened, people broke out in cheers. Olivia tried to explain the rules to me. The game went on for a long time, almost four hours, and it remained a mystery to me, but Olivia made it very interesting.

Cynthia, a young blond girl who worked as technician, also came to the lab with Edgar and Olivia. She liked parties, was dating a cameraman, and soon moved to Atlanta.

When Hunter's lab closed, Olivia moved back to cardiology, where she worked one floor above Yanhong.

WILLIAM

William, like Hunter, came from Britain. William was tall with a beard, often wearing bright and colorful shirts and slim. He exercised intensely and had a big lab. As a student, he sold wine in Harrods in London, which made me always think of him whenever I passed through London as I used to fly with British Airways. He graduated from Cambridge and went on to work for GlaxoSmithKline. His work concentrated on dopamine receptors and Hunter seemed to get along well with him. After two or three years, they were producing stuff. He married an opera singer, and they had two children. William sometimes stopped by to ask me about some text by Richard Wagner, since he knew I speak German. He had great interest in actually all of the arts. Giorgia and Claudio worked for him; so did Cooper, a technician; Aiden, another postdoc; Tamás, Joscha's friend from Hungary; Viginia, who generously gave bedding to the cleaning lady Afra; and Luan, who was studying for his MD and PhD and who had rescued the Polish cleaning guy on our fifth floor. Luan was about to get married, bought an apartment at downtown Watertower Place downtown, and always asked me about the prices of Rolex watches in Switzerland. An elderly Chinese technician whose name meant "Beauty of the East" also worked for William. I told Yanhong that she really looked beautiful. Yanhong laughingly went to tell her. Many others worked in William's lab, some for a limited time. Rumor had it that William priviledged a student from Vietnam, Grace, and Giorgia claimed piracy of authorship.

At one point, Angela, a postdoc from Germany, arrived in William's lab. In the beginning, Angela seemed very lonely and often sought me out as I speak her language. However, since I was usually in the middle of an experiment, I could not help her very often, which always made her quite angry. Nor did she understand when Hunter and I went out together. At times, we got the impression that Angela was looking for more than just friendship with me. Now it was Hunter who laughed as I would hurry around the floor locking doors behind me — and even, once, locking myself out. That was revenge. Luckily, Angela returned to Germany, taking a position in Berlin.

Later, William moved his entire lab to downtown.

AMELIA

Amelia was, as Yanhong put it, a "California girl." After earning her PhD in LA, she did a postdoc in Germany. Amelia was a sweet, warm-hearted lady who seemed to get along with everybody. Yanhong still remembers when Amelia was promoted from associate professor to tenured professor; Amelia ran a quite large lab on the fifth floor employing many Russians. Later, when Hunter became sick, she kept visiting him in the hospital and remained in phone contact with him. She was a faithful friend.

Amelia had several nice people in her lab. Among them was Meichen, a technician from China. Meichen had a baby boy. Her parents came to visit from China, and we took a couple of pictures together. When I showed the pictures to friends in Switzerland, they asked me, "Have you been working in China?" By chance, everybody except me was Chinese in the pictures.

Amelia was friends with Carol, who always improvised her lectures, whereas people like Hunter meticulously wrote out all the lectures for their entire course.

DAVID

David, a professor on the second floor, was also head of the department at some point. He was a nice Jewish guy, rather quiet and withdrawn, and was married having two daughters. I first met David when Hunter sent me to borrow some chemical, and David kindly not only gave me the chemical, but the keys to his lab so that I wouldn't need to wait in the future. That was because his lab unlike all others had been always locked. He had a very nice lab with a beautiful spacious office, where he would work often wearing a Hawaiian shirt. He earned his PhD from Harvard Medical School and worked with zebra fish. At one point, a colleague of mine who was looking for a position got in contact with David, whose name I had given to her, but she never made it to Chicago.

One time, David had a highway accident that completely destroyed his car, but he walked away without even a scratch. Another time, David's wife got really sick, and later, so did he, but he never revealed anything. Hunter once said that David was very theoretical, always thinking ahead about what questions to ask, while Hunter would focus on how to finish papers.

Benas, in charge of confocal microscopy, worked for David and, like his PI, was very quiet and not talkative. Like Vilma, he originally came from Lithuania. He was quite tall. Hunter said that Benas felt as though we were all below Benas' level. Whenever somebody asked Benas something, he would shrug his shoulders and give a short answer, but he would never go any further into it. It always seemed he was just too busy. Benas had earned his PhD from the UIC and worked as a postdoctoral fellow at the University of California. Since he came to the U of C in 1997, he has worked for David extensively. Whenever I would come to David's lab, I would always stop by to see Benas when I could to say hi and sometimes for short technical advice.

CHARLES

Charles, a professor of neurology from Taiwan, was married to Jane, who was also from Taiwan. She worked as a school psychologist, and they had two sons. Charles had a great sense of humor that spared no one. When one of his sons, a student in technology at the time, started dating a girl, Charles suggested that he should continue his studies in Germany, the "land of technology." But his son understood and said, "If you don't want me to date this girl, you can simply say so."

I met Charles through Shuang in Lynn's lab. Shuang had rented a room in Charles's house, and the next time I was looking for housing, Charles kindly took me in, too. Shuang and I lived upstairs. Her room was smaller than mine but had a built-in bathroom, while I had to use the common bathroom on that floor. The huge kitchen was downstairs in the basement, and Charles bought us a popcorn maker as special gift. At the entrance into the family's living room, which was kept separate from our part, Charles kept his old microscope from Taiwan. Their living room, one of the most beautiful I had ever seen, used Taiwanese style furniture and decorations.

Charles was an expert in good restaurants in and around Chicago. Often, on a Saturday, he would kindly invite me to a place currently recommended as the best restaurant in Chicagoland. We would drive up north for a long time, not returning before afternoon, and Charles would simply say that Hunter could wait. I once invited Hunter over for dinner. Though most people enjoy my cooking, Hunter did not, and Charles jokingly suggested I better should not invite him again.

Charles lived a few houses away from William, and they had been colleagues in the hospital before. Once, William asked Charles for something, and Charles went right to prepare it. But it took William three years to return it back to Charles, who teased William that he didn't expect to have so much time to prepare it.

Charles also knew Lynn very well and told us how she even had continued working highly pregnant. Charles just seemed to know everyone in the hospital.

When I moved out of Charles's house, he asked me to leave something so that I would return. I left my house shoes. The magic must have worked, as I always returned to see Charles and Jane.

Shuang worked for Lynn, whose lab I joined when I first came to Chicago. She and I were the same age and size, and we got along very well. Whenever we both had some time, we would go downtown with her friend Ning and another girl, who came from Brazil. One New Year's Eve, she invited me to a party hosted by an Indian postdoc and his family, very kind people. As I had a flight back to Europe later that night, we could only stay a short time, entertained by their two-year-old girl.

Shuang had a very strong will and became sort

of my second mental body guard, after Vilma, as she knew the PhD student I had come to work with in the first place. We would occasionally see Shuang's boyfriend, Gary, a graduate student in physics, who was also very nice. Later, they got married. While Shuang always wanted to return to China, Gary did not. After Shuang finished her PhD in Lynn's lab, she moved on to Columbia University in NY, where I visited her once. I accompanied her to a music lesson she took in order to learn a traditional Chinese music instrument, some kind of long board with strings, I had never seen. Later, she made her dream a reality and returned to China, and Gary had to follow.

Her friend Ning, who worked for Lynn, too, had a very sad story. She married very young, while still in her twentieth and a graduate student. One time, she and her husband, who studied in another town, went camping in Yellowstone Park, and by some accident her husband drowned. Ning had to go to the funeral in Shanghai before returning to school in Chicago. She published a very nice paper in the prestigious British journal *Nature*; after graduation, she went to Berkeley.

ABOUT 20 YEARS LATER

With the economy going bad in 2008, the department started to see severe cuts, leaving many labs with only few people working in them. Despite that, the U of C added new buildings on campus. After a wonderful Olympic-sized swimming pool, the Gerald Ratner Athletic Center, was built a few years ago, a new library looking like a greenhouse rose across from the Fermilab. This was donated by a couple who had once met at a library and got married. Down the same street, on Ellis, construction started on a huge new hospital for special cases. The COOP on 55th Street was transformed into a more expensive place, now called "Treasure Islands," which Yanhong simply called "Trash Islands," as she and Lilly never used it. The Social Science Research Building stopped serving the delicious Lebanese food that James and Lilly used to buy.

More importantly, the Surgical Department took over most of the fifth floor, leaving Amelia alone there.

Over the next few years, Hunter suffered three heart attacks and two strokes: a smaller one, which went unnoticed and a bigger one, which occurred occipital and left him with impaired vision. He could not read much anymore, he could not drive or hardly ride a bike and was restricted in the use of his left hand. That meant he could not perform or even follow up science any more, which he had loved and was forced to retire. This came as a complete shock to him. Lilly visited him and Evelyn a few times in Evanston. Considering his diagnosis, he was doing fine. Now, while awaiting a heart transplant, he had to accept getting a left ventricular assist device (LVAD) to control his blood pressure. He was quite annoyed by it, because it required a battery change every six hours, making travel very difficult. Despite of the risk, the device could show a failure on a flight, Hunter and Evelyn managed to travel a bit. Hunter was very annoyed by this machine, he was inseparable from and disappointed that he could no longer work in science. Now he needed something else to do to keep himself busy and he had started painting. He was too humble to show Lilly or anyone else his paintings, but Evelyn pulled up a few on the computer to show her. As his left hand was of limited use now, he would drop things unless he paid enough attention. A young guy from the neighborhood would go on walk with him and in rehab he had met a professor of history, he would talk to and compare his limitations with him. Hunter suffered sleeping problems and he became quite anxious. At the skype sessions with Lilly he could hardly stay concentrated for an hour, then he became restless and he had to go. He had considerably changed and he had become the nicest, caring friend of everyone, commenting on Merkel as Lilly lived in Germany at that time and no longer being the restless scientist he used to be. His musical tastes also changed, and we were quite surprised to learn that Hunter now enjoyed Michael Jackson. One of his favorites, "Smooth Criminal," he called "quite compelling and a work of a genius." And again, he talked about his memory, which he called fragmented with the impairment of his short-term memory. However, his long-term memory we found remarkably fine and in attempt to bring his memory back Lilly asked him to tell his story. From the notes she took at that time we could write about his past and more.

Hunter's friends Giorgia and Claudio went to Philadelphia and fell out of contact with Hunter. But Hunter stays in touch with other friends. Tali and Lior still see Hunter from time to time for dinner. Their youngest son is a smart kid and goes to a small school, Williams College. He is great at math and may pursue a scientific career.

Tali works with Lior in molecular genetics, both having a lab of about six people.

Heinz, a graduate student, remains a good friend of Hunter.

Recently, Hunter asked Lilly to look up a guy he used to work with from Dr. Greengard's lab at Yale, Uwe. She found Uwe working as a professor in Mainz, Germany, where she, too, recently worked in the hospital. So Lilly met Uwe a couple of times for coffee at the campus of Mainz University. She gave him Hunter's contact information, and Uwe dropped him a nice note. Since then, Hunter has gotten back in touch with Uwe and his wife.

By now Bill occupied only a small lab with only one student on the first floor, where Edward used to work. Lilly did not meet Bill the last time she went back, as he was out of town for a late Thanksgiving with his sons. Now he does more teaching about signal transduction pathways to graduate students. His wife Alison has a photography studio, and both of them travel and write articles for a travel magazine.

Both of his former students, Pjotr and Chang, have gotten married by now, and we wondered how Chang impressed his wife without Lilly's violin.

Scarlett left Bill's lab to finish her PhD in Grenoble, France; she married a biochemistry postdoc from Britain. They bought a house in Grenoble and lived there with their two little sons until they decided to move back to Britain. Her husband decided not to work as a postdoc anymore and instead became a schoolteacher, while Scarlett considered working for the British Navy.

Joscha still works at the Argonne National Laboratory and was still "changing diapers," until recently— this time for his sister-in-law's children. We teased him that he was by now probably the most professional diaper changer, but all that matters was Joscha's happiness. When Yanhong once suggested that he should play the lottery more often, Joscha replied that he has already gotten everything he wished for. One unfulfilled dream remained though: to work at Harvard Medical School. Joscha had repeatedly applied for a post there and even got accepted, but he could not take the position due to family bonds. Maybe when his children and the children of his sister-in-law are grown up and move, Joscha can fulfill even this dream, if he still wants to go there after so many years.

His son Dominik decided he had poured enough gels for his life and went on to study architecture.

After Yanhong passed all of her medical exams, got a green card, and even worked as an assistant professor, she applied for a residency program so she could finally work as a physician in the US. She had sent out several applications, and when she received a letter of acceptance, she couldn't remember where that place was. Excited, she called out to Tanner "some island!" Poor Tanner's heart must have sunk, since it could have been Hawaii, which is very far away. When Yanhong looked it up, it turned out to be Staten Island, NY.

So, through the National Resident Matching Program, she entered the NYMC (New York Medical College) residency program in pediatrics. All of us have asked her: "Why pediatrics?" She replied: "First of all, you decide on an organ. Mine was always the kidney. In kidney disease, you see a lot of old people, and if you have the chance to see young ones, why not head for children?" She became a PGY 1 (Post graduate year one) pediatrics resident. By now, she has completed NICU (Neonatal Intensive Care Unit), PICU (Pediatric Intensive Care Unit), newborn nursery and emergency room rotations, and has gained significant ability in patient care and management. She plans to use her expertise from years of research in nephrology to specialize in pediatric kidney diseases. Her residency program was very tough. When Lilly visited her in 2012 in Staten Island, she was either working or on call, which meant usually she had to work, too. She really did not have much time to prepare for her board exams. Yanhong used her holidays to participate in the board exam reviews in several states. She passed everything and settled down in Houston as a children's kidney specialist. Once she passed the board exam, like all others in the US she must do it again every ten years. This includes collecting credits every year by joining conferences. In Europe, where Lilly lives, doctors are spared from that. Once having passed the board exam they are good to work until they get retired. In addition, this program, which was installed in the seventies, appears much more difficult than the requirements of other countries. But Yanhong didn't stop here. She is just graduating from pediatric nephrology fellowship right now and accepted a position as a clinical associate professor in NY starting in December. As Liudmila kept saying all the years, "Yanhong could be a role model for all of us!"

We asked Yanhong why she didn't practice medicine in Canada, as she holds a Canadian passport. She replied, "Medicine in Canada is socialized. Doctors there cannot work overtime, moonlight, or continue to work at any age to make more money." It is the same in Europe.

As always, Yanhong and Lilly continue to have very long open discussions whenever Lilly comes to visit, though these occasions have become rare. These last late into the night and cover just everything from life strategies to the advantages and disadvantages of marriage, and future plans. Yanhong still keeps Lilly's plants, which she kindly offers to return to her each time they talk. In November 2016 Yanhong attended the ASN conference in Chicago and stepped by to see Bill, Amelia and Sam.

Yanhong's friend Tanner still lives in Toronto. His mom in Connecticut has passed away, and he and his brother sold her house, the house where he grew up. Yanhong's brother is moving to San Diego, as her sister-in-law just accepted an offer to be a full professor and chairwoman at the Department of computer science, UCSD. They are moving to California in August 2017. And Yanhong still talks to Prof. Bernstein.

Liudmila was visiting Moscow the last time Lilly went to Chicago. She worked at Northwestern and lived on 61st Street and Ingleside with Dima for several years until they moved to Kentucky. By then, her cat, Persik, "the king", she was proud of, had died. She has gotten a new companion for her other cat, Archibald. Yanhong and Lilly called her from Yanhong's new home in Staten Island during Easter 2012. Poor Liudmila had undergone another surgery by then but has recovered nicely. The three plan to reunite in Chicago at some point.

After travelling around for school, earning her MFA (Master of Fine Arts) in Experimental Animation from the California Institute of the Arts) and moving several times for work e.g., to Portland, Oregon, for a cartoon movie, Naomi finally settled back at her grandmother's home in San Francisco. In Israel, she married an American teacher, and they have a son and two daughters. Naomi works as a teacher, too, and animates for a video company as well. She writes: "In the end, I think of my years in Chicago fondly. Just this morning, I considered applying for another job there".

After Vilma moved to Houston and completed a one-year residency program for physician assistants, she stayed there working in the GU Medical Oncology Department of the Anderson Center ever since. Her parents left Chicago and settled in Houston, as well. In 2009, Vilma married an associate professor in industrial engineering at the University of Houston of Korean descent, who just got awarded. They have a girl and a boy. Sadly her Dad has passed away recently in 2017.

Betty started working at UIC, because they offered a better retirement plan, but recently lost her job and was looking for a new position. But she meanwhile found the love of her life, and the two live together.

Pam moved from Boston at the Harvard Medical School to a job at Bio-Rad.

Laura got married and was happily living in Shanghai with her family.

Raj continued working as a postdoc at the DKFZ in Heidelberg, Germany, and was looking into options for becoming a professor. Raj and Lilly met occasionally, and they even made plans to talk with Hunter via Skype together. And Raj made his dream become true: He became a professor, and like Hunter said before, it is all about publish or perish. Raj hopes to get tenured soon and is working in Britain, where Hunter comes from.

Katarina still lives off campus up north and was not planning a career in science, which would involve writing grants. She preferred a safe job to leave her with more time for her children and grandchildren. When Hunter had to close his lab, Katarina moved to another building and became CORE director with Hunter's help.

Jingle must have had her baby and has probably graduated from Northwestern University.

Theodore passed away in 2012 and was buried in Israel.

Edgar still lives off campus — not far from Hunter, actually.

Olivia, still in cardiology, had her 30-year anniversary at the U of C, for which she was invited for cheese and wine. We all joked about how much longer she needed to keep working to be awarded a full dinner.

William meanwhile got retired and keeps on publishing. He is the author of articles and even books, can be seen on youtube and is passionate about the environment and the life of animals.

Amelia was the only professor still working on our fifth floor, and still runs her own lab. She underwent knee surgery and had lost weight when we met her last time. She was waiting to get the other knee done as well. Much to the amusement of Yanhong, she introduced Lilly last time she came to visit to one of her postdocs, Kontantin, by saying, "Kontantin, here comes your Christmas present!" - leaving the guy expecting wonders. It turned out that Kontantin was not only curious about working for the US government as Lilly had done before, but he had been working for a professor in Germany, who is the brother-in-law of one of Lilly's friends.

David still worked on the second floor and was hoping to live in Paris, France upon retirement. He and Lilly talked about his daughters. The younger one joined the State Department corps as a teacher overseas. He was running the department in Chicago. In 2012, Benas appeared on a webinar about microscopy, which Hunter and Lilly talked about afterwards.

Charles wanted to sell his house. However, due to the bad economy, many houses in Hyde Park were for sale at that time, so he decided to stay. His wife Jane retired, too, and they are occasionally babysitting for their grandchildren. The last time we met, his first words, reflecting his sense of humor, were: "Lilly! You gained weight!" He added that he meant it as a compliment, as she had always been too skinny. (Lilly had gained maybe a grand total of two kilograms over the last twenty years.)

Shuang now runs her own lab in fertilization in Beijing.

Lynn, her previous PI in Chicago, had moved as well and was no longer at the U of C. Raj and Lilly even joined a lecture of her, she presented at the DKFZ in Heidelberg, where they have been both working at that time. Just a few months after Raj sent us a sad note that she had passed away, not believing they had just seen her.

After so many years, Dan and Lilly did not go on. Since he had become captain money had become very important to him while she had other priorities. They had a beautiful, exciting time, meeting in places all over the world and living in Israel, and we really loved his great sense of humor. But it was not meant to last and they finally went different ways. Career-wise, Lilly had followed Yanhong's model and chose her "favorite organ", which always was the brain. Lilly did her specialization in toxicology and neurosurgery and worked in epilepsy, brain tumors and trauma. She returned to Basel, where she had been once working for the Basel Zeus.

By the time Lilly settled in Switzerland, she had lived in Japan, Israel, Italy, Germany and the US, learning other languages, which she loves and never thinks to be good enough at.

While others are focused in one field, Lilly always loved variety. Beside her work in medicine she had spent considerable time in rain forests every year. There she would collect material for toxicology and see patients. In addition, she also holds a diploma in zoology, which allows her to study wild animals and fight for their protection. She obtained a flying license while working at the U of C and she could move more independently in those places.

Though she could not be hired by the FBI, she held a side job with the US government for almost twenty years. She also served as the official translator for the minister of foreign affairs of the Philippines. And she went on writing literature and has received several awards. She also still plays her violin and piano and she has plans for the next academic title.

In May 2014 she had another bike accident and this time she was severely injured when a teenage boy rushed to school in the morning, while she was on her way to work. She spent about six months in a wheelchair and had to relearn walking for one year. That experience left her with high attention when she walks outside today and she would never use her bike again in traffic.

In March, 6th 2017 Lilly lost her last close family member in an accident.

Lilly keeps seeing Radu and Liesel, who live in Basel, too. Radu completed his PhD in chemistry and went on to become a High School teacher. Liesel had a risky heart valve replacement a few years earlier and was getting older, restricted in things she could do but was doing fine. Her cousin Ursula had undergone open heart surgery in Amsterdam and was restricted as well.

Most of us went to different places, but what stays is our memory of the time in Chicago. However, several of us have become very sick and the saying "after 30 disease looks for man" has sadly become all too true.

On Lilly's visits, she mostly returned to Chicago during the winter, which made Joscha ask once why they were always taking pictures freezing in the snow. Since an Illinois driver's license expires on one's birthday, and her birthday comes in December, winter always seems like a good time to come. Joscha suggested that she changes her birthday.

P.S.:

Hunter and Lilly stayed in touch via Skype until recently in June 16th in 2017, when his LVAD finally indeed failed and left him without a pulse – leaving us praying for him while he was in a coma. On the same day, Liesel was admitted to the ICU with a lung inflammation and in urgent need of another valve replacement, while the newspapers reported about former Germany's chancellor Helmut Kohl's passing. Only a few days later, on June 22nd we learnt that Hunter's life supporting machines had been stopped and that he had passed away. Vilma and Lilly already had suffered great losses this year. Reminded on Tanner's words once how fragile life is Xiaoyan and I decided that we already waited too long with this book and it should get published now. It is sad that Hunter is no longer able to read it. Hunter died age 71. We lost a great scientist and our best friend.

It is the story of where all of us met and many of us still keep fond memories. Parents of future students, who may consider the U of C for their children, should keep in mind: Bad things happen everywhere; but there is only one magic place: For us it was the U of C, Chicago's Hyde Park, Chicago.

GLOSSARY OF NAMES- THE "WHO IS WHO"

All names have been changed except of the ones of Nobel laureates.

Abigail: postdoc in Tali's lab
Adam: son of Joscha
Adrian: professor on the third floor
Afra: cleaning lady
Aiden: postdoc in William's lab
Alfredo: friend of Hunter
Alison: wife of Bill
Amelia: professor on the fifth floor
Andy: director of the Kentucky Reptile zoo
Angela: visiting scientist and postdoc in William's lab
Angelo: son of Giorgia and Claudio in Williams lab
April: janitor in Lilly's building

Ariel: Department chairman in Israel
Ava: professor in the hospital

Bassem: Naomi's ex boyfriend
Benas: professor in David's lab
Betty: graduate student in Hunter's lab
Bodhi: professor and Department chairman
Brooke: postdoc in Tali's lab
Bruce: professor and Department chairman, PI of Carol

Carlos: professor and Department chairman
Carol: professor, friend of Amelia
Chang: student in Bill's lab
Charles: neurology professor in the hospital
Claudio: postdoc in William's lab, husband of Giorgia
Cooper: technician in William's lab
Cynthia: technician in Theorore's lab

Dan: Lilly's boyfriend
David: professor on the second floor
Debra: friend of Lilly, who works at the US State Department
Diego: friend of Lilly, professor in Mexico
Dima: husband of Liudmila
Dolores: ex girlfriend of Hunter and postdoc in Sam's lab

Dominik: son of Joscha
Dr. Carlsson: Nobelprize laureate 2000
Dr. Greengard: Nobelprize laureate 2000
Dr. Kandel: Nobelprize laureate

Edgar: postdoc, cardiologist in Theodore's lab
Eddie: grandson of Edward
Edward: professor on the second floor
Elijah: son of nurse in the hospital, member of the Chicago Gospel choir
Emma: contact family in Toronto for Yanhong
Erich: friend/partner of Liesel
Evelyn: later wife of Hunter

Gad: business partner of Ilana
Gary: Shuang's husband
Gina: technician in Hunter's lab
Giorgia: postdoc in William's lab, wife of Claudio
Grace: student in William's lab
Gus: graduate student and fellow of Hunter in Israel

Harper: assistant professor on campus
Hao: Yahong's former classmate
Heather: assistant of Andy of the Kentucky Reptile zoo
Heinz: graduate student (of or with Hunter), friend of Hunter

Hunter: professor on the fifth floor

Ian: son of Edward
Ido: former PI of Hunter in Israel
Ilana: ex wife of Hunter

James: professor of social science on campus
Jane: wife of Charles
Jenny: Vilma's friend and roommate
Jingle: student in Hunter's lab
Joyce: Edward's secretary
Joscha: graduate student in Bill's lab
Juri: Liudmila's son in Moskow

Ka: brother of Yanhong
Kason: son of Ka, nephew of Yanhong
Kang: professor on campus
Katarina: postdoc in Hunter's lab
Kelly: girlfriend and later wife of James
Konstantin: postdoc in Amelia's lab

Laura: student in Hunter's lab
Leela: part of the study group at Kaplan with Yanhong
Leesha: secretary on the third floor
Leonardo: friend of Giorgia and Claudio
Leroy: secretary on the third floor

Liesel: friend of Lilly in Basel
Lilly: visiting scientist and postdoc in Hunter's lab
Lior: husband of Tali, professor in the hospital
Liudmila: postdoc in Bill's lab
Luan: graduate student in William's lab
Lynn: professor on the second floor

Matthew: graduate student of Hunter
Marc: receptionist in Lilly's building
Meichen: technician in Amelias's lab
Michelle: Edward's wife
Moshe: PI of Hunter in Israel
Miyu: wife of postdoc on campus

Naomi: technician in Bill's lab
Nelson: friend of Lilly, works for the US State Department
Ning: graduate student in Lynn's lab

Olivia: technician in Theodore's lab
Owen: fellow of Hunter in Yale

Pam: graduate student in Hunter's lab
Peng: professor on the second floor
Pjotr: student in Bill's lab
Professor Bernstein: former PI of Yanhong in Toronto
Professor Hinze: one of Lilly's professors in Basel

Professor Zak: one of Lilly's professors in Basel
Put: technician in Tali's lab

Radu: Lilly's best friend in Basel
Raj: technician in Hunter's lab
Roman: colleague of James in social sciences on campus
Ron: student in Hunter's lab
Russell: professor at Yale

Sam: professor in the hospital
Sanja: postdoc on campus
Santosh: professor in Knapp building
Scarlett: Lilly's friend, visiting scientist in Bill's lab
Seo-yun: technician in Bill's lab
Shanti: part of the study group at Kaplan with Yanhong
Shuang: graduate student in Lynn's lab

Tali: professor on the third floor
Tamás: postdoc in William's lab
Tanner: partner of Yanhong
TC: neighbor of Lilly
Theodore: professor in the hospital
Tom: friend of Sanja

Udo: postdoc in Prof Bernstein's lab in Toronto
Ursula: cousin of Liesel, Lilly's friend
Uwe: fellow of Hunter in Yale

Vilma: student in Hunter's lab
Viktor: professor in Toronto, PI of Yanhong
Virginia: student in William's lab

Wang: Yanhong's former classmate
Warren: professor on campus
William: professor on the fifth floor

Yanhong: postdoc and later Assistant professor in Bill's lab

Zachary: professor in Toronto, PI of Yanhong
Zorica: later wife of Joscha

www.ingramcontent.com/pod-product-compliance
Lightning Source LLC
Chambersburg PA
CBHW020637220526
45464CB00001B/181